Decision Support Systems for Multisectoral Tropical Lake Management in Indonesia

Klaus Glenk • Ivana Yuniarti
Rosetyati Utami • Atiqotun Fitriyah
Gunawan Pratama Yoga
Editors

Decision Support Systems for Multisectoral Tropical Lake Management in Indonesia

Nature and Anthropocene in Lake Singkarak

Springer

Editors
Klaus Glenk
School of Natural and Social Sciences
Scotland's Rural College (SRUC)
Sustainable Ecosystems Team
Edinburgh, UK

Rosetyati Utami
Research Center for Limnology and Water Resources
Indonesian Research and Innovation Agency (BRIN)
Bogor, Indonesia

Gunawan Pratama Yoga
Research Center for Limnology and Water Resources
Indonesian Research and Innovation Agency (BRIN)
Bogor, Indonesia

Ivana Yuniarti
Research Center for Environmental Archaeology, Marine Archaeology, and Sustainable Culture
National Research and Innovation Agency (BRIN)
Bogor, Indonesia

Atiqotun Fitriyah
Research Center for Limnology and Water Resources
Indonesian Research and Innovation Agency (BRIN)
Bogor, Indonesia

ISBN 978-3-032-11267-5 ISBN 978-3-032-11268-2 (eBook)
https://doi.org/10.1007/978-3-032-11268-2

This Springer imprint is published by the registered company Springer Nature Switzerland AG
The registered company address is: Gewerbestrasse 11, 6330 Cham, Switzerland

Foreword

Water is the most encompassing global-scale integrator, connecting aquatic and terrestrial ecosystems and the atmosphere in a continuing cycle of use, replenishment, and re-use. Adequate supplies of freshwater are an absolute requirement for all life and socioeconomic development on our planet, also being of cultural and/or religious significance to some societies. At the same time, however, less than one percent of all the freshwater on the Earth's surface is in liquid form, with more than 90% of this quantity existing in lakes, wetlands, marshes, and floodplains (hereafter collectively referred to as lakes). There are lakes on every continent, providing a wider range of life-supporting ecosystem services than other aquatic systems, being a major resource for addressing human health needs, poverty alleviation, food production, recreation, tourism, hydropower production, and preservation of biodiversity integrity.

Lakes, whether natural or manmade, are also among the most readily accessible freshwater sources available to humanity. Indeed, this accessibility is a major reason that lakes are so vulnerable to the negative impacts of anthropogenic activities. To this end, some have described lakes as "The Mirrors of the Earth," highlighting their role as the ultimate integrators and indicators of the negative impacts of human activities within, and even outside, their catchment areas.

The unique defining characteristics of lakes, including an integrating nature to inputs, complex in-lake dynamics, and long water retention times, also make their assessment and management a complex challenge, particularly in regard to ensuring sustainable use of their many life-supporting ecosystem services. Indeed, we have made great advances in developing freshwater resources in many places. At the same time, however, our ability to manage lakes in a sustainable manner has unfortunately lagged far behind our ability to exploit their myriad of ecosystem services. This is attributable to the fact that sustainable lake management requires consideration of a wide range of complex and interacting socioeconomic, institutional, policy, financial, and political challenges. Addressing these challenges requires a comprehensive and participatory management approach, ideally within the context of an integrated lake basin management platform.

To facilitate addressing these lake-related assessment and management challenges, this book highlights the rationale and value of applying user-friendly decision-support systems (DSS) to obtain comprehensive information to facilitate the lake management decision-making processes. It discusses the development of such DSS, as well as the lessons learned from DSS application to Lake Singkarak in Indonesia. It also highlights the need for, and value of, interdisciplinary studies in applying DSS to lakes exhibiting diverse anthropogenic and natural pressures contributing to their degradation. Describing the experiences of DSS application in the tropical climatic zone of Indonesia is also informative from a limnology perspective, noting much of the limnological knowledge currently used to assess and manage lakes around the world is still largely based on studies done in the temperate regions of our planet.

Readers should find this book both informative and useful in evaluating the interplaying governance systems underlying the use and sustainability of the study lake, as well as highlighting the challenges in the use of DSS processes. Accordingly, it will be beneficial to a range of readers, including lake practitioners, researchers, decision-makers, managers, and civil society, in addressing the many natural and anthropogenic challenges facing the sustainability of lakes and the range of life-supporting ecosystem services they provide to humanity.

Chair, Scientific Committee
International Lake Environment Committee,
San Antonio, TX, USA

Walter Rast

Book Summary

The digital era has transformed decision-making across the globe. However, the adoption of decision support systems (DSS) in lake management—especially in Global South countries—remains limited. Lakes in these regions, predominantly located in the tropics, undergo massive anthropogenic and natural pressures, often leading to intense degradation under uncertain conditions. These circumstances are further exacerbated by the lack of data and scientific evidence to aid the decision-making process.

Indonesia, home to over 800 tropical lakes, offers a valuable case study for understanding the situation. The large demography, rapidly growing economy, relatively low human resources development index, and increasing frequency of extreme weather have simultaneously pressurized its lake ecosystems. Therefore, eutrophication, sedimentation, and water-use conflicts are the country's common stories. Yet, evidence-based approaches to decision-making for lake planning and monitoring systems remain scarce. Hence, developing a user-friendly decision support system (DSS) is imperative in these regions.

This book emphasizes the development, early implementation, and lessons learned from such DSS. The DSS integrates multiple domain-specific models—including hydrology, water quality, fisheries, species distribution, GIS, and economics—demonstrating the power of interdisciplinary approaches and bridging the gap between academic research and practical use. Additionally, the science, governance systems, and challenges in developing the DSS are fascinating, offering valuable lessons for lake-related scientists and decision-makers. This suggests that the book can be both informative and helpful for practitioners and scholars, especially in the Global South.

Chapter 1 offers an overview of the book's context and its application. The context covers what the book discusses, general knowledge of lake management in the Global South and Indonesia, a brief description of the decision support system (DSS) project in Lake Singkarak, the scope of the book and the DSS, the knowledge gaps that the book and project address, and the novelty of the project.

Chapter 2 critically examines natural backgrounds and how human activities around Lake Singkarak's catchment area impact its water quality and ability to support economic growth. It provides context for the study area and discusses current and future projections of the causal relationships between these factors. It also investigates the institutional challenges, such as local and legal aspects, of lake management to offer governance insights and show how they influence decision-making and the lake's current state.

Chapter 3 elucidates how the foundation framework of the DSS was built. We constructed an analytical framework for the Ecosystem-Based Management–Drivers-Pressures-States-Ecosystem Services-Responses (EBM-DPSER). The framework provides robust scientific fundamentals and legitimacy for the DSS since it was constructed using stakeholder and expert-based knowledge. It was able to portray the dynamics of the lake ecosystems and the complex interactions of their constituent components (both abiotic and biotic components and the socio-economic systems).

Chapter 4 provides a glimpse of the DSS, comprising its features and the covered management problems. This chapter elucidates the inclusion of processes to maintain the legitimacy of the framework and the data, and the efforts to ensure data equity and the incorporation of uncertainties. It also reflects on the limitations and scope of the DSS and the implications of the power of data, and possibly obscuring (further) uncertainties in the data itself.

Chapter 5 simulates the development of hydrological models (SWAT and QWET) to predict the impacts of several land use management scenarios on improving the lake's water quality. These silo models are two of the backbones of the DSS. Meanwhile, Chapter 6 focuses on the fisheries and habitat-environment models as substantial parts of the DSS and how they are connected to the DSS to predict the future impacts of several management scenarios on the lake's fish stocks. It also discusses the proposed conservation efforts of an endemic fish habituated in the lake and the future prediction of its impacts on the fisheries. Chapter 7 describes current tourism and related policies at various government levels. It also simulates a tourism model specification connected with the environmental sub-model to predict the impacts of the management scenarios on tourist visitation.

Chapter 8 presents how both principal outputs of the DSS (cost-benefit analysis [CBA] and Sustainability Index [SI]) are generated. It summarizes the economic valuation process used to calculate the CBA of each management scenario. The economic value includes economic valuation of the reduction/increment of turbidity, marginal rice field flooding, habitat conservation, and macroplastic waste management. Meanwhile, Chapter 9 summarizes the information technology used to integrate the silo models into a user-friendly DSS. It also presents examples of the DSS results to evaluate several management scenarios. Finally, the last chapter (Chapter 10) contemplates the lessons learned from the development of the DSS and our plan to apply it in the study area. Further, it shows how the DSS emerged in the policy-making process in the study area as a valuable lesson for future research and management in Indonesia and the Global South. It also reflects our experience of the effect of integrating participatory and evidence-based decision-making and elaborates on data legitimacy and equity.

Contents

About the Editors

Klaus Glenk is an environmental and resource economics professor and leads the Sustainable Ecosystems Team at Scotland's Rural College (SRUC). He has 20 years of experience in applied cost-benefit analysis of environmental policy interventions, including climate change mitigation and adaptation in agriculture, biodiversity conservation, and water management. Prof. Glenk's research has been published in leading disciplinary and interdisciplinary journals and focuses on the micro-analysis of consumer and household decisions. He has extensive experience in the methodological development and application of stated preference methods, particularly choice experiments and cost-benefit analysis. Another strand of work includes investigating resource use decisions, technology choice, and contract design for incentive-based land management schemes such as payments for ecosystem services of farm households in Europe and the Global South. Recent work also covers issues of demand for (food) products with sustainability and health claims. Prof. Glenk has been involved in research projects covering a broad range of environmental, food, and land management issues within Europe (Spain, Germany, UK), Asia (China, Indonesia, Myanmar), Sub-Saharan Africa (Kenya, Uganda), and South America (Brazil).

Ivana Yuniarti, Ph.D. is a researcher in the Research Center for Environmental Archaeology, Marine Archaeology, and Sustainable Culture at the Republic of Indonesia National Research and Innovation Agency. She received her Ph.D. from the University of Edinburgh and has 15 years of experience in interdisciplinary approaches to tropical lake management. Her works range from ecological modeling to ecological-economic methods (cost side and stated preference methods). She has published in both disciplinary and multidisciplinary journals and actively reviews publications in lake management and ecosystem services assessment. Her current projects include the development of lake decision support systems in Indonesia. Her previous research was mostly in Southeast Asian Countries (i.e., Thailand and Indonesia).

Rosetyati Utami, Ph.D. is a researcher at the Research Center for Limnology and Water Resources, National Research and Innovation Agency (BRIN), Indonesia. She holds a Ph.D. in Environmental Engineering from Institut Teknologi Bandung (ITB), with research focusing on environmental risk assessment of emerging pollutants in aquatic systems, particularly in river basins. Her scientific contributions span passive sampling development, ecological and human health risk modeling, and contaminant prioritization—mainly addressing pesticides, pharmaceuticals, and heavy metals. She has led and collaborated on multiple national and international projects, including partnerships with Radboud University and Deltares (The Netherlands), and funding from Kurita Water and Environment Foundation (Japan). Her editorial interests lie in advancing interdisciplinary studies on water quality, ecotoxicology, and risk governance in data-limited contexts, particularly in Southeast Asia. She brings a strong commitment to scientific rigor, collaboration, and applied environmental solutions. She actively reviews for international journals and supports science-policy integration through community-based environmental work and mentoring early-career researchers.

Atiqotun Fitriyah, Ph.D. is a researcher at the Research Center for Limnology and Water Resources, National Research and Innovation Agency (BRIN), Indonesia, specializing in hydrology and remote sensing for water management. She holds a Ph.D. in Environmental and Agricultural Engineering from Tokyo University of Agriculture and Technology. Her work encompasses remote sensing-based assessment for the water environment, SWAT-based hydrological modeling in Southeast Asian River basins, and water balance analysis in mountainous agricultural systems. Her current projects include monitoring water allocation through remote sensing in Bali and evaluating green infrastructure for disaster risk reduction in Banten. Her multidisciplinary approach integrates fieldwork, satellite data, and remote sensing techniques to provide actionable solutions for global water issues.

Gunawan Pratama Yoga, Ph.D. is a senior researcher at the Research Center for Limnology and Water Resources, National Research and Innovation Agency (BRIN) in Indonesia. He has 25 years of experience in freshwater ecotoxicological research. His works range from the acute effect of toxicants to their ecological and health risks. His current research is modeling and prioritizing active pharmaceutical ingredients (APIs) in Indonesia. His previous study mainly dealt with mercury pollution in some artisanal gold mining in Indonesia.

Abbreviations

AHP	Analytical hierarchy process
AWD	Automatic watershed development
BMP	Best management practice
BOD	Biological Oxygen Demand
CA	Cellular automata
CBA	Cost-benefit analysis
CHRIPS	Climate Hazard Group Infrared Precipitation with Station
CLD	Causal loop diagram
COD	Chemical oxygen demand
CPI	Consumer Price Index
CR	Consistency ratio
GUI	Graphical user interface
JSON	JavaScript Object Notation
DEM	Digital elevation modelling
DO	Dissolved oxygen
DSS	Decision-support systems
DST	Decision support tool
DPSIR	Drivers-Pressures-States-Impacts-Responses
EEA	European Environmental Agency
EBM-DPSER	Ecosystem-Based Management–Drivers-Pressures-States-Ecosystem Services-Responses
EBM	Evidence-based management
ERD	Entity relationship diagram
ES	Ecosystem services
ESA	Environmental system analysis
FAO	Food and Agriculture Organization
GIS	Geographical Information Systems
HRC	Habitat replacement cost
IAD	Institutional Analysis and Development
IMF	International Monetary Fund
IT	Information technology

HRU	Hydrology response units
MAPE	Mean absolute percentage error
MC	Markov chain
MSY	Maximum sustainability yield
NPV	Net present value
NSE	Nash-Sutcliffe efficiency
PBM	Participatory-based management
PLN	Perusahaan Listrik Negara
PSDA-PU	Pengairan dan Sumber Daya Air Pekerjaan Umum
SD	System dynamics
SDG	Sustainable Development Goal
SFD	Stock-flow diagram
SGA	Stock growth algorithm
SI	Sustainability Index
SRTM	Shuttle radar topography mission
SSPM	Schaefer Surplus Production Model
SWAT	Soil and Water Assessment Tool
SWOT	Strength Weakness Opportunities Threats
TDS	Total dissolved solids
TN	Total nitrogen
TP	Total phosphorus
TS	Target strength
TSI	Trophic State Index
TSS	Total suspended solid
UN	United Nation
USA	United States of America
USGS	United States Geological Survey
WCA	Water catchment area
WET	Water Ecosystems Tool
WTP	Willingness to pay

Chapter 1
Introduction: Context and Knowledge Gaps

Ivana Yuniarti, Klaus Glenk, Rosetyati Utami, Atiqotun Fitriyah, Gunawan Pratama Yoga, Hidayat, and Aldiano Rahmadya

Abstract The chapter provides the book's and the decision support system (DSS) contexts in the Global South and Indonesia's economic and political settings. It also elucidates the general context of lake management on a global and national scale, while also highlighting the use of the ecosystem services framework in shaping the decision support systems. The discussions are pointed to the question that should be answered to substantially improve lake management in Indonesia and in the Global South, which includes the principal actors, actions to take, and how to implement them. Further, it underlines the knowledge gaps and novelties of the book. It ends with a brief guideline on how to use the book for the readers. Overall, it contemplates the roles of the book and the DSS in Lake Singkarak management to offer solutions to water challenges.

Keywords Lake · Global south · Decision support system · Lake management

1.1 The Context in the Global South and Indonesia

The eco-socio-political division of nations into the Global North/ South and similar dichotomies has been challenged but is still in use among official institutions such as the United Nation (UN) and the International Monetary Fund (IMF) [1, 2]. The

I. Yuniarti (✉)
Research Centre for Environmental Archaeology, Maritime Archaeology, and Cultural Sustainability, National Research and Innovation Agency (BRIN), Cibinong, Indonesia
e-mail: ivana.yuniarti@brin.go.id

K. Glenk
Sustainable Ecosystems Team, School of Natural and Social Sciences, Scotland's Rural College (SRUC), Edinburgh, UK

R. Utami · A. Fitriyah · G. P. Yoga · Hidayat · A. Rahmadya
Research Centre for Limnology and Water Resources, National Research and Innovation Agency (BRIN), Cibinong, Indonesia

K. Glenk et al. (eds.), *Decision Support Systems for Multisectoral Tropical Lake Management in Indonesia*, https://doi.org/10.1007/978-3-032-11268-2_1

Global South countries, which are commonly characterized by lower socio-economic living standards, comprise the regions of Africa, Latin America, the Caribbean, Asia (excluding Israel, Japan, and South Korea), and Oceania (except Australia and New Zealand) [3]. These nations, contributing to about 44% of the global export value, showed a strong post-pandemic economic recovery [4]. As emerging and competitive economies, they often exhibit a much wider and imbalanced distribution of education and technological skills, income, and poverty, leading to higher resource competitiveness, urbanization rates, and pressures on natural resource exploitation [4, 5]. Combined with poor resource governance, the rapidly emerging economy of these countries also generates further drawbacks including elevated water demands that trigger unsustainable water withdrawals, water crises and conflicts, and swift degradation of water sources [6–8]. These challenges necessitate a focus on research acceleration, capacity development, and technological interventions to manifest the concept of sustainable development as guided by Sustainable Development Goals (SDGs) [6–9].

Turning our attention to Indonesia, a burgeoning economic giant in Southeast Asia, the country is home to diverse tropical ecosystems ranging from enormous tropical rainforests to mountainous ecosystems, as well as coastal and oceanic areas [10, 11]. In addition to the common challenges faced by countries of the Global South, Indonesia encounters additional multi-faceted challenges, including eco-political turbulence and natural resource degradation at large scale [12, 13]. Nevertheless, the country strives despite such situations while struggling to maintain the availability and quality of its natural resources [14, 15].

As one of the most populous countries in the world, Indonesia ranks 55th in the water scarcity index in 2024 and is a highly water-stressed country in 2040 [16, 17]. Moreover, it holds the 75th rank in the sustainability index among 166 evaluated countries [18], highlighting the urgency of research and technological interventions. Along with the development of information technology (IT), applying evidence-based decision-making becomes much more straightforward with multi-sectoral decision support systems (DSS).

This book serves as a recollection of knowledge behind the development of a novel DSS in Lake Singkarak, Indonesia, offering solutions to water challenges. The focus of the book is on lake ecosystems, which play a crucial role in providing clean water sources as well as various ecosystem services that yet remain relatively understudied concerning their importance for a sustainability transition in the Global South [19].

1.2 Concepts of Sustainable Lake Management

Sustainability or sustainable development has resonated since 1980 [20]. The term is connoted with the improvement and sustenance of ecological, economic, and social systems for human development [21, 22]. In the context of lake management, it is commonly translated as the inclusion and sustenance of these three pillars by accommodating various stakeholders' needs with consideration of the ecological capacity of the ecosystems [9].

Despite regular citations, there are no clear and agreed-upon guidelines on what a *sustained lake* is. Some studies associate sustainable lake management with active stakeholder participation, reliable data, well-managed conflicts, and a circular economy, and most of them relate it to better water quality [23, 24]. The traditional engineering/limnology approaches even narrow the sustainable lake management goal to water clarity [23], which may only be appropriate in certain contexts [24, 25].

The global movement of sustainable lake management directs the countries' governments to align it with integrated water basin management, particularly focusing on the protection and use of ecosystems with particular emphasis on their water quality, erosion, sedimentation, and biodiversity [26]. The suggested approaches include participatory management, conservation, climate adaptation, good governance, and technology (ibid). While this sounds comprehensive, in general, there are no clear directions and technical interventions for the national government on how to conserve lakes effectively.

We argue that it is essential to tailor the criteria for the success of lake management, as they serve as the fundamental frame for prioritizing management actions. Some factors that affect the criteria selections include the economic level of the country, the multi-sectoral nature of lake uses, climate change, political agendas, and the availability of funds. In the advanced economic countries, the success of lake management is mostly defined by the lake's health [27]. Meanwhile, in the Global South, heavy emphasis is placed on the adequacy of provisioning ecosystem services to support both local and national economies [28]. Economic status is highly correlated with the fourth factor, the availability of funds. There have been cases where, in an emerging country, environmental restoration is directed according to the agenda of the donor country while undermining the local recipient's needs [29, 30]. At the same time, the governance type and the occurrence of polycentric governance in lake management also complicate the establishment of the success criteria [31, 32].

In this book, we mainstream the concept of ecosystem services as a framework for the development of suitable criteria that define sustainable lake management in Indonesia, as the concept of ecosystem services comprises a broad inclusion of the understanding of human and nature complex relationships [33]. The criteria include:

1. Higher economic value of ecosystem services (ES) with fewer trade-offs of the ESs.
2. More cost-effective management solutions.
3. Achievement of common understanding among the stakeholders.
4. Increased water quality (better water transparency, lower sediment, and nutrient loading).
5. Good and transparent governance (i.e., participatory, occurrence of stakeholders' group, clear rules, and sanctioning).

These criteria will frame the chapters in the book, which are included in the built Decision Support System (DSS). As prior information, we do not include the mitigation of the effects of climate change in the indicators due to data limitations for building forecasting models.

1.3 The Context of Lake Management in Indonesia

The lakes in Indonesia are used for multiple purposes, including fisheries and aquaculture, hydropower generation, tourism, clean water sources, and the sinkholes of domestic and municipal waste [27]. Geologically, the ecosystems were formed from either volcanic eruptions (crater lakes) or tectonic activities. As common-pool resources, the ecosystems are subject to a high level of subtractability, making it costly to exclude other potential beneficiaries [28]. This attribute aligning with the boost of the human population generally implicates further consequences such as establishing an open-access regime, which creates complex negative externalities (e.g., stakeholder conflicts, eutrophication, pollution, and mass fish kills) [27, 32, 34–36].

Our rapid review distinguished that among 23 lakes in Indonesia selected based on the availability of recent reports of their trophic status (Appendix 1), two were categorized as oligotrophic, six as mesotrophic, six were identified as eutrophic, and four were classified as hypereutrophic. Four lakes are in transitional status, two in oligotrophic to mesotrophic status, and three in eutrophic to hypereutrophic status.

This situation generated concern among national decision-makers, which resulted in the declaration of 15 national priority lakes in 2009, followed by the second list in 2014 (Fig. 1.1). These lakes are the target of the national restoration project, which formally obligates the central government to provide substantial funds for their restoration with the support of the local government [33]. The declaration was followed up with various formal events, such as the declaration of collective actions and issuance of presidential regulation: Perpres no.60/2021 [37, 38].

The "how" queries generate deeper contemplation on:

1. How are the resources and ecosystems used and managed?
2. How are participatory-based management (PBM) and/or evidence-based management (EBM) applied?

Nevertheless, almost two decades after the 2009 declaration, there have been no significant improvements in the lakes' water quality, as indicated by the latest report on three major lakes. The most recent report from Lake Tempe in South Sulawesi indicated that physicochemical parameters, including turbidity, Total Suspended Solid (TSS), Biological Oxygen Demand (BOD), nitrate, and phosphate, have exceeded the water quality standards set by the Government of the Republic of Indonesia regulation No. 82/2001. This has led to a decrease in the quality and status of Lake Tempe, classified as moderately polluted [39]. In the case of Lake Maninjau, a study revealed that fish production and feed quota from fish culture with floating net cages have surpassed the capacity of Lake Maninjau's water pollution loads. This disturbance has adversely affected the aquatic ecosystem, resulting

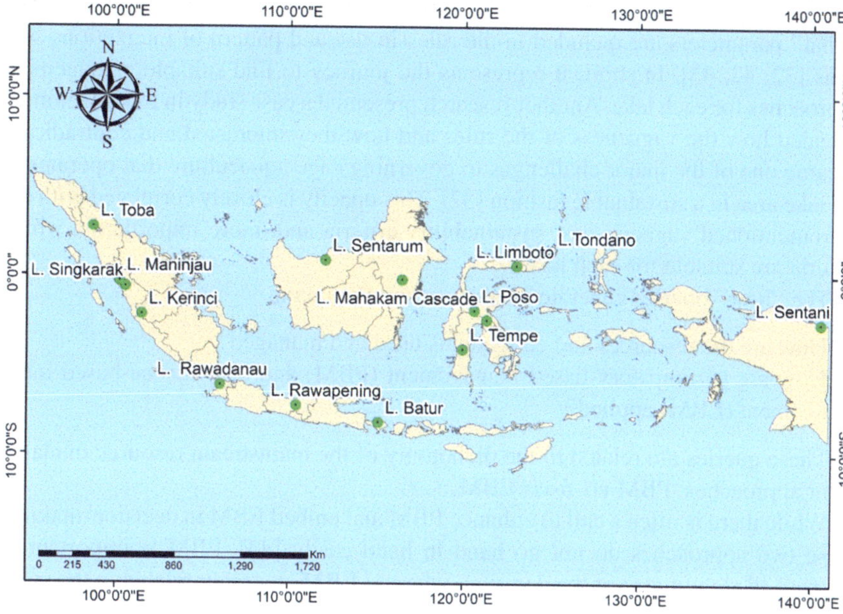

Fig. 1.1 National priority lakes of the Republic of Indonesia

in the water quality status of Lake Maninjau being categorized as heavily polluted, with a trophic status of hypereutrophic [40]. Similarly, nutrient accumulation in the form of nitrogen and phosphorus in Lake Toba has posed challenges. Another study assessed the trophic state index (TSI) based on chlorophyll content, Secchi disk depth or water transparency, and total nitrogen and phosphorus content, categorizing Lake Toba as a meso-oligo-eutrophic lake [41]. The lake exhibits consistent mesotrophic characteristics due to abundant nutrient availability at all sampling stations. Compared to a previous study in 2009, the trophic state of Lake Toba has shifted from oligotrophic to mesotrophic, requiring special attention to prevent eutrophication through excessive phosphorus enrichment in the future. These situations illustrate the current general conditions of other major lakes in the country.

From the managerial perspective, there is a non-negligible confusion about the "who, what, and how" related questions concerning the lakes. The "who" questions are related to vagueness about the actors involved in the management, such as the property rights owners, the principal manager of the lakes, and the main funders (cf. [39, 40]). The "who" factor becomes one of the bottleneck problems in Indonesian lake management. As for other ecosystem types, such as rivers, specific local authorities (councils) manage each river catchment area (e.g., Badan Otoritas Sungai Asahan). Nonetheless, no such council can manage the stakeholders' interplay for the Indonesian lakes (cf. [9]).

Meanwhile, the "what" inquiries are intercorrelated with the ambiguities of the rules applied in constitutional, collective, and operational levels, as well as the goals and success indicators. In Ostrom's Institutional and Development Framework, the "what" parameters are included in the rules-in-use and pattern of interactions sections [32, 42, 43]. In short, it represents the journey to find suitable management approaches for each lake. Another research presented a case study in Lake Maninjau revealed how the vagueness of the rules and how they interacted and contradicted became one of the major challenges to governing cage aquaculture that operates in the lake area in a sustainable fashion [32]. This opacity is closely correlated with the aforementioned vagueness of sustainability criteria and, more importantly, which criteria are suitable for each lake.

The "how" queries generate deeper contemplation on:

1. How are the resources and ecosystems used and managed?
2. How are participatory-based management (PBM) and or evidence-based management (EBM) applied?

These queries are related to the dichotomy of the mainstream resource management approaches: PBM *vis-à-vis* EBM.

While there is often a call to enhance PBM and embed EBM in decision-making, these two approaches do not go hand in hand easily [44]. PBM is important to increase the legitimacy of the decision, whereas EBM is closely related to the credibility and adequacy of scientific evidence and thus of the managerial options [44, 45]. PBM is more adaptive, transparent, and responsive to public opinions, but it is more susceptible to politicization [44, 46]. In contrast, EBM, which represents the technocracy approach to the decision-making process, is more exclusive and requires rigorous and systematic academic approaches [47]. Thus, it is challenged by the quality of research culture, decision-maker capability, and the absence of guidance on what data and literature can be considered evidence (ibid).

Elsewhere, at the international level (i.e., in the EU), there have been efforts to integrate PBM and EBM, those two advancements, but still confusion and difficulties persist, and are [44, 46]. In Indonesia, where the basic culture is communal, the original decision-making and resource management approach are participatory-based, as shown by the occurrence of enduring community-based management, especially in forest and coastal areas [48–52]. However, this is not generally the case for lake management, with limited records on community-based lake management. Moreover, aligning with the country's polycentric governance and the political turbulence, lake PBM is a rare and scattered story; however, there are cases, they are but barely documented in the context of the primary lakes, in most cases found for small lakes or reservoirs (e.g., Malahayu Reservoir, Urban Lake Rawa Besar) [53].

Further, despite continuing efforts to endorse lake EBM, the associated processes are still far from optimal. While research institutions and universities have collected abundant data and suggested recommendations, most are written in scientific journals, theses, and reports using academic terms. More importantly, the available scientific evidence is patchy, scattered, and not presented to align with the decision-maker's needs. Furthermore, scientific evidence collected by academics usually requires longer gathering and analyzing, which may not align with decision-making needs. However, the national decision-makers (who allocate funds) require easy-to-read information (for example in the form of indices) to monitor and plan the funding distributions. At the local level, the government requires fast and measurable information to adapt to its annual targets, commonly called "professional norms." To fulfill the evidence sections in management planning, most governmental agencies rely on rapid assessments conducted by consultants, whose results are not well-tested and evaluated in the scientific community. This circumstance is a common problem at the science-policy interface, as hinted at in the study "Does research information meet the needs of stakeholders?" explored evidence in the global management of invasive species [54–56]. The above challenges hamper the endorsement of lake EBM in Indonesia, contributing to further disorientation of Indonesia's lake management, which at present is neither well-planned, participatory-based, nor evidence-based.

Overall, there is confusion about how the lakes should best be managed due. To overcome such confusion, systematic solutions such as an online database are advised by academics, as proposed by [57]. While an online database is essential, in the practical decision-making process, the science and policy interface can be facilitated via the development of standalone models and a more advanced tool, like a decision support system, to integrate the models.

1.4 Knowledge Gaps and Novelty

Concluding the discussions, lake management in Indonesia as an illustrative case for other Global South countries is neither participatory nor evidence-based. Although there have been attempts to follow either of these approaches or to combine both, there is not enough evidence of their success. Therefore, there is a pressing need for accelerated research to inform which management approach is suitable for each lake and should thus be endorsed.

Growing concern about this situation motivates the development of the DSS presented in this volume. The tool is a breakthrough effort to make evidence-based decision-making on sustainable lake management easier and simpler. The novelty of this DSS lies in its role as a hub that connects various siloed models using advancements in IT. Moreover, applying the participatory approach in developing the conceptual framework (Ecosystem-Based Management-Drivers-Pressures-States-Ecosystem Services-Responses or NEBM-DPSER framework), as explained in Chap. 3, is a novel method for developing tropical lake DSS. This approach is our dedicated contribution to the fundamental challenge of aligning both participatory and evidence-based decision-making as contemplated by [58].

1.5 How to Use the Book

The book reflects our work on Lake Singkarak. It presents the uniqueness of the lake, which may generate general readers' interest. It particularly elucidates the development of the DSS in the lake, a tectonic lake in West Sumatra, which represents natural resources governance in an Islamic and matrilineal culture. Therefore, as the book may interest specific readers such as scientists and lake decision-makers, general readers such as students and journalists are also likely to enjoy it. However, although acquiring a background in IT makes it easier to read Chap. 9, we ensure that the chapter can also be understood by people without IT knowledge.

The book can be read as separate chapters, as each chapter summarizes each silo model. However, it is advised to follow from Chap. 2 to the last chapter to understand the concept and context of the science behind the DSS. Chapter 2 provides an overview of the lakes and their human dimensions. Meanwhile, Chap. 3 explains the foundation for the entire book and highlights how the participatory approach is integrated into the DSS. From Chap. 4 to the last chapter, the silo models are introduced, and it is explained how they are connected using the IT hub. The final chapter offers a synthesis of lessons learned through developing and applying the DSS in the study areas.

We hope that the book can provide general guidance for developing DSS in other lake areas in Indonesia and other Global South nations. Thus, the summary of lessons learned offers valuable information for scientists and decision-makers contributing to the development of sustainable lake management strategies.

References

1. Viera Valencia LF, Garcia Giraldo D. Introduction. The Global South and/in the Global North: interdisciplinary investigations. Global South. 2019;8:1–15.
2. Rambaldi C. A review of the development divide between global north and south through a Foucauldian perspective. Development Studies Research. 2022;9:67–9. https://doi.org/10.1080/21665095.2022.2042348.
3. The United Nations Conference on Trade and Development (UNCTAD). Country classification. 2023. Available: https://unctadstat.unctad.org/EN/Classifications.html. Retrieved on 11 January 2024.
4. United Nations. Handbook of statistics 2022. New York: United Nations Publications; 2022. Available: https://unctad.org/system/files/official-document/tdstat47_en.pdf
5. Thornton JA, Thornton JA, Rast W, Ryding SO. The role of socio-economic determinants in lake management in developing countries. Lake Reserv Manag. 1991;7:115–20. https://doi.org/10.1080/07438149109354260.
6. United Nations. The United Nations world water development report 2021: valuing water. Water politics. Paris: The United Nations Educational, Scientific and Cultural Organization; 2021. https://doi.org/10.4324/9780429453571-2.
7. Boretti A, Rosa L. Reassessing the projections of the world water development report. NPJ Clean Water. 2019;2 https://doi.org/10.1038/s41545-019-0039-9.
8. Sjöstrand K. Water for sustainable development. Nat Water. 2023;1:568–72. https://doi.org/10.1038/s44221-023-00108-2.
9. Yuniarti I, Barnes C, Glenk K, McVittie A. Towards sustainable Lake ecosystem-based management– lessons learned from interdisciplinary research of cage aquaculture management in Lake Maninjau. In: Triyanti A, Djalante R, Marfai MA, Indrawan M, Nurhidayah L, editors. Environmental governance in Indonesia. Springer; 2023.

10. HYSH N, Nurfatriani F, Indrajaya Y, Yuwati TW, Ekawati S, Salminah M, et al. Mainstreaming ecosystem services from Indonesia's remaining forests. Sustainability (Switzerland). 2022;14 https://doi.org/10.3390/su141912124.
11. Nurhati IS, Murdiyarso D. National Strategy for Mangrove Ecosystem Management A reference for the conservation and rehabilitation of coastal areas to achieve Sustainable Development Goals and Low Carbon Development. Working Paper 14. Bogor; 2023. Available: https://www.cifor-icraf.org/publications/pdf_files/WPapers/CIFOR-ICRAF-WP-14.pdf
12. Castañeda JP, Raina L, Rexhepi D. Synthesis report: natural capital accounts and policy in Indonesia wealth accounting and the valuation of ecosystem services (WAVES) program in Indonesia. Washington, DC; 2019.
13. Bayu Mogana Putra BMP, Ridwan Adnan Erlangga. Analysis of Indonesian political economic policies in National Economic Recovery due to Global Pandemic Turbulence. J Integ Int Relation. 2022;7:95–108. https://doi.org/10.15642/jiir.2022.7.2.95-108.
14. Firdaus N, Supriatna, Supriatna J. Ecosystem services research trends in Indonesia: a bibliometric analysis. Biodiversitas. 2022;23:1105–17. https://doi.org/10.13057/biodiv/d230255.
15. Anas T, Hill H, Narjoko D, Putra CT. The Indonesian economy in turbulent times. Bull Indones Econ Stud. 2022;58:241–71. https://doi.org/10.1080/00074918.2022.2133344.
16. Gassert F, Reig P, Luo T, Maddocks A. Aqueduct country and river basin rankings: a weighted aggregation of spatially distinct hydrological indicators. Washington, DC: World Resources Institute; 2013. Available: wri.org/publication/aqueduct-country-river-basin-rankings
17. World Population Review. Countries with Water Scarcity 2024. In: Countries with Water Scarcity 2024 [Internet]. 2024. Available: https://worldpopulationreview.com/country-rankings/countries-with-water-scarcity
18. Sachs JD, Lafortune G, Fuller G, Drumm E. Sustainable Development Report 2023. Dublin: Sustainable Development Solutions Network; 2023. https://doi.org/10.1007/978-3-642-28036-8_101577.
19. Lovgren S. Rivers and lakes are the most degraded ecosystems in the world. Can we save them ? National geographic [Internet]. 2021; [cited 30 Sep 2023]. Available: https://www.nationalgeographic.com/environment/article/rivers-and-lakes-are-most-degraded-ecosystems-in-world-can-we-save-them?awc=19533_1640258848_2df30856bee79f013d2b4415a7b644b0
20. Klarin T. The concept of sustainable development: from its beginning to the contemporary issues. Zagreb Int Rev Econ Bus. 2018;21:67–94. https://doi.org/10.2478/zireb-2018-0005.
21. Purvis B, Mao Y, Robinson D. Three pillars of sustainability: in search of conceptual origins. Sustain Sci. 2019;14:681–95. https://doi.org/10.1007/s11625-018-0627-5.
22. Mensah J. Sustainable development: meaning, history, principles, pillars, and implications for human action: literature review. Cogent Soc Sci. 2019;5:1653531. https://doi.org/10.1080/23311886.2019.1653531.
23. Baker LA, Newman RM. Managing the biological, economic, and social aspects of sustainability of lake ecosystems. Comprehensive water quality and purification. Elsevier; 2013. p. 76–86. https://doi.org/10.1016/B978-0-12-382182-9.00063-3.
24. Tammeorg O, Chorus I, Spears B, Nõges P, Nürnberg GK, Tammeorg P, et al. Sustainable lake restoration: from challenges to solutions. WIREs Water. 2023:1–26. https://doi.org/10.1002/wat2.1689.
25. Dokulill MT, Donabaum K, Katrin Teubner. Restoration and lake management. In: Dokulill·MT, Donabaum K, Teubner K, editors. The Alte Donau: successful restoration and sustainable management, aquatic ecology series 10. Cham: Springer; 2018. p. 407. Available: http://www.springer.com/series/5637
26. UNEP. Resolution adopted by the United Nations Environment Assembly on 2 March 2022. Nairobi; 2022. p. 4–6.
27. Novianti R, Yuniarti I, Ajie GS, Setiawan F, Triharyuni S, Afandi AY. System analysis and research priorities towards sustainable management of Lake Singkarak, Indonesia. In progress. In: IOP conference series: earth and environmental science. IOP Publishing; 2023.

28. Ostrom E. Institutional analysis and common-pool resources. In: Ostrom E, Gardner R, Walker J, Agrawal A, Blomquist W, Schlager E TS, editor. Rules, games, and common-pool resources. Michigan: The University of Michigan; 1994. p. 38.
29. Lewis TL. Environmental aid: driven by recipient need or donor interests? Soc Sci Q. 2009;84:144–1. https://doi.org/10.1111/1540-6237.8401009-i1.
30. Weiler F, Klöck C, Dornan M. Vulnerability, good governance, or donor interests? The allocation of aid for climate change adaptation. World Dev. 2018;104:65–77. https://doi.org/10.1016/j.worlddev.2017.11.001.
31. Cianci-Gaskill JA, Klug JL, Merrell KC, Millar EE, Wain DJ, Kramer L, et al. A lake management framework for global application: monitoring, restoring, and protecting lakes through community engagement. Lake Reserv Manag. 2024;40:66–92. https://doi.org/10.1080/10402381.2023.2299868.
32. Yuniarti I, Barnes C, Glenk K, Sutrisno. Challenges for the development of environmentally sustainable cage culture farming in Lake Maninjau, Indonesia: an institutional perspective. Ecosyst People. 2021;17:248–63. https://doi.org/10.1080/26395916.2021.1921848.
33. Martin-Ortega J, Jorda-Capdevila D'd, Glenk K, Holstead KL. What defines ecosystem services-based approaches? In: Martin-Ortega J, Ferrier RC, Gordon IJ, Khan S, editors. Water ecosystem services: a global perspective. Cambridge: The United Nations Educational, Scientific and Cultural Organization (UNESCO), Paris, and Cambridge University Press; 2015.
34. Yuniarti I, Glenk K, McVittie A, Nomosatryo S, Triwisesa E, Suryono T, et al. An application of Bayesian Belief Networks to assess management scenarios for aquaculture in a complex tropical lake system in Indonesia. PLoS One. 2021;16:1–23. https://doi.org/10.1371/journal.pone.0250365.
35. Kharisma A. Sampah & Penangkap Ikan Resahkan Wali Nagari Salingka Danau Singkarak. Detik.com. 2022. Available: Atta Kharisma.
36. Simatupang KH, Arlan A, Felicia A. Urgensi Penyelamatan Danau Prioritas di Indonesia: Studi Kasus Upaya dan Komitmen Penyelamatan Lingkungan Hidup Danau Toba. J Tour Creativ. 2022;6:216. https://doi.org/10.19184/jtc.v6i3.34993.
37. Central Government of The Republic of Indonesia. Peraturan Presiden (Perpres) Nomor 60 Tahun 2021 tentang Penyelamatan Danau Prioritas Nasional. 2021 p. 14. Available: https://peraturan.bpk.go.id/Home/Details/171165/perpres-no-60-tahun-2021
38. PPN/Bappenas, Kementerian Komunikasi dan Informatika. Aksi Kolektif untuk Tujuan Bersama: Pengelolaan 15 Danau Prioritas Nasional. 2017.
39. Yani A, Rohman F, Suarsini E. Water quality and pollution index of Lake Tempe in South Sulawesi, Indonesia. Pollut Res. 2019;38. Available: https://www.researchgate.net/publication/340933793
40. Syandri H, Azrita A, Mardiah A. Water quality status and pollution waste load from floating net cages at Maninjau Lake, west Sumatera Indonesia. IOP Conf Ser: Earth Environ Sci. 2020; https://doi.org/10.1088/1755-1315/430/1/012031.
41. Barus TA, Wahyuningsih H, Hartanto A. Water quality and trophic status of Lake Toba, North Sumatra, Indonesia. Hydrobiological Journal. 2022;58:34–43. https://doi.org/10.1615/HYDROBJ.V58.I2.30.
42. Ostrom E. Rules, games, and common-pool resources. The University of Michigan Press; 1994.
43. Ostrom E. Governing the commons: the evolution of institutions for collective action. Cambridge University Press; 1990.
44. Bunea A, Chrisp J. Reconciling participatory and evidence-based policymaking in the EU Better Regulation policy: mission (im)possible? J Eur Integr. 2023;45:729–50. https://doi.org/10.1080/07036337.2022.2144848.
45. Cash DW, Clark WC, Alcock F, Dickson NM, Eckley N, Guston DH, et al. Knowledge systems for sustainable development. Proc Natl Acad Sci U S A. 2003;100:8086–91.
46. Rauh C. Clear messages to the European public? The language of European Commission press releases 1985–2020. J Eur Integr. 2023;45:683–701. https://doi.org/10.1080/07036337.2022.2134860.

47. Baba VV, HakemZadeh F. Toward a theory of evidence based decision making. Manag Decis. 2012;50:832–67. https://doi.org/10.1108/00251741211227546.
48. Astuti EW, Hidayat A, Nurrochmat DR. Community forest scheme: measuring impact in livelihood case study Lombok Tengah Regency, West Nusa Tenggara Province. Jurnal Manajemen Hutan Tropika. 2020;26:52–8. https://doi.org/10.7226/jtfm.26.1.52.
49. Wahyu A, Suharjito D, Darusman D, Syaufina L. The development of community-based forest management in Indonesia and its contribution to community welfare and forest condition. IOP Conf Ser Earth Environ Sci. 2020;528 https://doi.org/10.1088/1755-1315/528/1/012037.
50. Widyastini T, Hadi Dharmawan A. Efektivitas Awig-Awig dalam Pengaturan Kehidupan Masyarakat Nelayan di Pantai Kedonganan Bali. Jurnal Sosiologi Pedesaan. 2013;1:37–51. Available: http://repository.ipb.ac.id/handle/123456789/66071
51. Damastuti E, van Wesenbeeck BK, Leemans R, de Groot RS, Silvius MJ. Effectiveness of community-based mangrove management for coastal protection: a case study from Central Java, Indonesia. Ocean Coast Manag. 2023;238:106498. https://doi.org/10.1016/j.ocecoaman.2023.106498.
52. Rakatama A, Pandit R. Reviewing social forestry schemes in Indonesia: opportunities and challenges. Forest Policy Econ. 2020;111:102052.
53. Saragi PA, Budhi Soesilo TE, Abidin CA. Review on community-based strategies for improving the quality of Lake waters (case study: Urban Lake Rawa Besar, Depok, West Java, Indonesia). IOP Conf Ser Mater Sci Eng. 2020:012029. https://doi.org/10.1088/1757-899X/725/1/012029.
54. Bayliss HR, Wilcox A, Stewart GB, Randall NP. Does research information meet the needs of stakeholders? Exploring evidence selection in the global management of invasive species. Evidence Policy. 2012;8:37–56. https://doi.org/10.1332/174426412X620128.
55. Cook CN, Hockings M, Carter RW. Conservation in the dark? The information used to support management decisions. Front Ecol Environ. 2010;8:181–8. https://doi.org/10.1890/090020.
56. Bainbridge I. How can ecologists make conservation policy more evidence based? Ideas and examples from a devolved perspective. J Appl Ecol. 2014;51:1153–8. https://doi.org/10.1111/1365-2664.12294.
57. Sutherland WJ, Pullin AS, Dolman PM, Knight TM. The need for evidence-based conservation. Trends Ecol Evol. 2004;19:305–8. https://doi.org/10.1016/j.tree.2004.03.018.
58. Brownlie WJ, Woods HJ, Waters KE, Moore AL, Bruce AM, Olszewska JP, et al. Freshwater science for the benefit of society: a perspective from early career researchers. Inland Waters. 2017;7:227–35. https://doi.org/10.1080/20442041.2017.1326793.

Chapter 2
Lake Singkarak: Natural History and People

Ivana Yuniarti, Sulung Nomosatryo, Unggul Handoko, Aiman Ibrahim, Iwan Ridwansyah, Tri Suryono, Nina Hermayani Sadi, Gunawan Pratama Yoga, Radityo Bambang Wicaksono, Syahroma Husni Nasution, Ahmad Yusuf Afandi, and Ira Akhdiana

Abstract Written in timeline style: past, present, and future, the chapter reflects on Lake Singkarak's natural and human dimensions to provide the background knowledge of the lake and the importance of developing a multisectoral decision support system. It elucidates the "wicked" problems encountered by the lake managers and their implications for current and future lake conditions. Understanding the evidence from the sediment coring, we highlight the lake's natural tendency to undergo eutrophication, which directs the lake's ecosystem responses towards anthropogenic and climatic stressors. Further, the effects of people's resource use patterns and current governance are also elaborated. At the end of the chapter, we summarize several important indicators that reflect the current state, indicating the need for the decision support system as an evidence-based management aid tool.

Keywords Lake Singkarak · Lake management · Decision support system · Eutrophication · Ecosystem · Environmental wicked problems

---

I. Yuniarti (✉)
Research Centre for Environmental Archaeology, Maritime Archaeology, and Cultural Sustainability, National Research and Innovation Agency (BRIN), Cibinong, Indonesia
e-mail: ivana.yuniarti@brin.go.id

S. Nomosatryo · A. Ibrahim · I. Ridwansyah · T. Suryono · N. H. Sadi · G. P. Yoga · S. H. Nasution · A. Y. Afandi · I. Akhdiana
Research Centre for Limnology and Water Resources, National Research and Innovation Agency (BRIN), Cibinong, Indonesia

U. Handoko
Directorate of Laboratory Management, Research Facilities, and Science and Technology Park, National Research and Innovation Agency (BRIN), Cibinong, Indonesia

R. B. Wicaksono
Department of Resources and Environmental Economics, Faculty of Economy and Management, IPB University, Dramaga, Indonesia

K. Glenk et al. (eds.), *Decision Support Systems for Multisectoral Tropical Lake Management in Indonesia*, https://doi.org/10.1007/978-3-032-11268-2_2

2.1 Chapter Introduction

Lake Singkarak, a shining mirror of rich natural history and culture, is a perfect study site for a typical Global South tropical system. It is located in Tanah Datar and Solok District in West Sumatra Province and exhibits multiple resources used to support the locals' livelihoods and the island's energy generators.

We must look beyond its current state and dwell on its history to understand its current condition. The lake's history offers lessons for how it responds to human intervention; thus, it may mirror the future of the ecosystem and the people. This chapter guides readers through time while reflecting on the intertwined fate of people and nature.

2.2 Lake as Lentic Ecosystem

Before elaborating on Lake Singkarak's condition, we highlight that analyzing lake ecosystems differs from analyzing river ecosystems. As a lentic water body with an ecosystem of relatively still water, a lake exhibits unique physical, biological, and chemical characteristics [1]. Furthermore, lakes are defined as naturally occurring bodies containing standing water, which allows for depth features [1]. Depth is a critical component that affects lake ecosystems, influencing significant aspects of the lake system, such as light penetration, vertical temperature profiles, dissolved oxygen profiles, mixing, and decomposition processes [1]. Therefore, these parameters are important for understanding and analyzing lake ecosystems in contrast to river ecosystems, where such parameters may be less varied, if at all. Eventually, when determining lake conditions, academics rely on the trophic status level or the Delphi Method [2].

2.3 The Daunting Past

Firstly, we trace back to the lake formation history, as this becomes one of the critical characteristics, which affect the lake's system and responses to external drivers. It has long been believed that Lake Singkarak has a tecto-volcanic origin [1]. However, this view has been challenged, and it has been suggested that the lake is formed by graben damming by Mountain Marapi-Singgalang and Tandikat volcanic materials [2].

Secondly, we refer to the sedimentary study by [3] since it portrays both geological and sedimentary analysis of Lake Singkarak. Their work shows that the lake has a naturally low sedimentation rate (~2.0 mm per year) unless natural hazards such as earthquakes create high debris avalanches. Further, they also identified that even in the past, the lake was prone to experiencing algae blooming associated with organic materials input from volcanic activities. This phenomenon was also strengthened by long-term water transparency [4]. Using satellite data, the authors observed that during the 1980s–1990s, the lake's Secchi depth was relatively low (on average below 2 m), with gradually increasing trends up to the 2020s.

Interestingly, our analysis of the trends of precipitation rate and occurrence illustrates that both climatic parameters are trending upward (Fig. 2.1). We analyze that

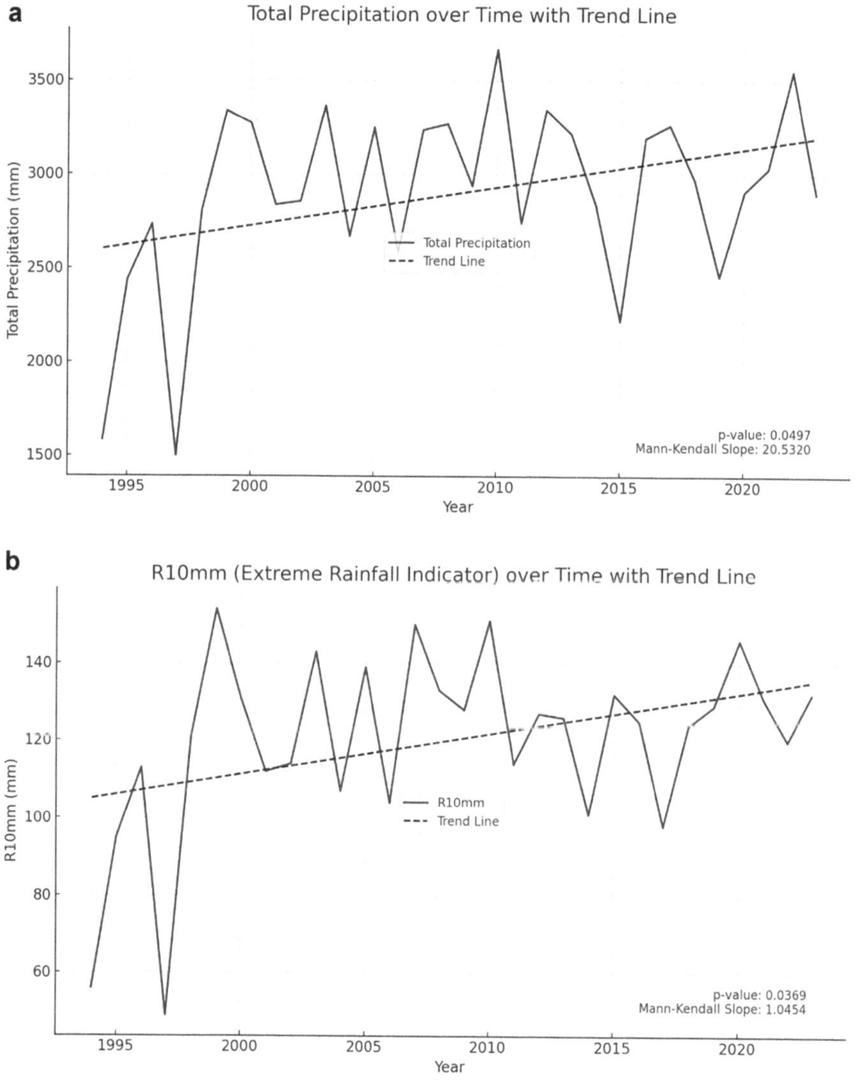

Fig. 2.1 Trends of (**a**) precipitation rate, (**b**) extreme precipitation rate. (Source: Authors' simulation using RClimDex based on Climate Hazard Group Infrared Precipitation with Station (CHRIPS): Rainfall Estimates from Rain Gauge and Satellite Observations from 1994 to 2023)

Table 2.1 SWOT analysis of Lake Singkarak's tourism

Strengths	**Weaknesses**
Natural beauty and unique ecosystem	Limited tourism infrastructure
Strong local culture	Lack of promotion and tourism information
Ecotourism potential	Unclear management guidelines
Opportunities	**Threats**
Nature and cultural tourism trends	Environmental degradation (pollution, waste)
Government support (event *Tour de* Singkarak)	Land conversion
Local economic development program	

Source: Authors' analysis based on a primary survey in April 2024

from 1994 to 2023, the precipitation rate increased almost 1.4 times (Fig. 2.1a). Further, the extreme precipitation rate was elevated by about 40% (Fig. 2.1b). This condition contributes to the higher chance of flooding and erosion potential.

The remaining question is why water transparency elevates with higher precipitation and extreme precipitation rates. The answer lies in the success of the replantation program by the Ministry of Forestry. Our analysis of land use changes and hydrological conditions in the lake (details in Chap. 4) shows that forest coverage almost doubled from 1990 to 2020, resulting in surface flow dwindling by about 2.4% and during the same period, impacting the decline of erosion and sedimentation rates (Table 2.1).

2.4 Resource Use Pattern and Anthropogenic Activities in the Catchment Area

2.4.1 Lake's Catchment Area and Number of Population

Lake Singkarak provides various ecosystem services for locals and other stakeholders in the adjacent area. Here, we analyze how people use the lake resources, especially fishers. First, we present the resource system from the catchment area perspective and illustrate the spatial extent of anthropogenic activities that can affect Lake Singkarak's ecosystem (Fig. 2.2). Two major watersheds expand from the upstream of the Sumpur River in Tanah Datar District and Sumani River in Solok District (Fig. 2.2). In total, the lake's catchment area approaches 102.7 ha [5], which encompasses Padang Panjang and Solok cities, four sub-districts in Tanah Datar District, and four sub-districts in Solok District (Table 2.2).

There has been a large-scale settlement in the lake's catchment area over the past two decades, indicated by the increase of population (Fig. 2.3). The increase has been particularly occurring in the urban areas of Solok and Padang Panjang cities, where the population has increased by about 33% and 20%, respectively, over 16 years (Fig. 2.3). Meanwhile, in other *nagaries* (customary villages) surrounding the lake, the population numbers remain steady (Figs. 2.3 and 2.4).

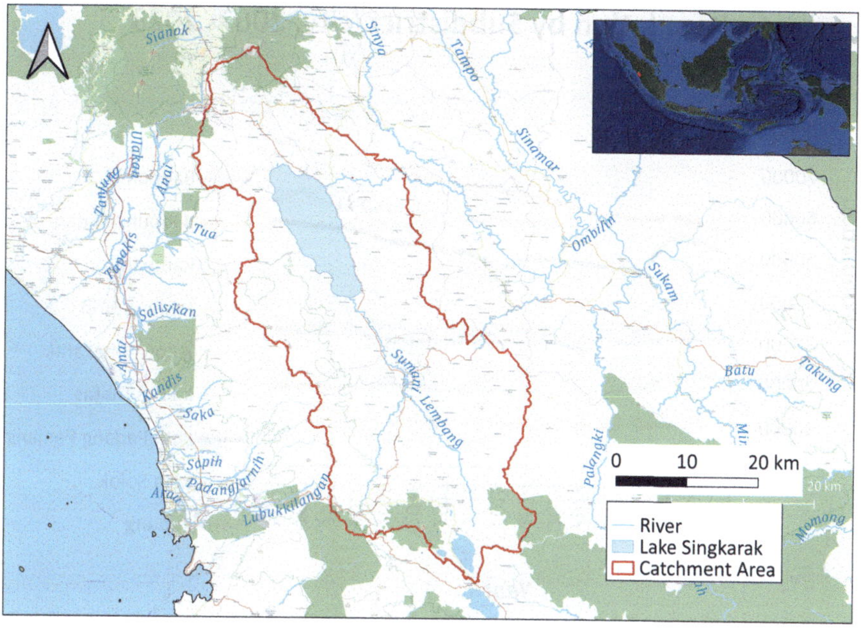

Fig. 2.2 Lake Singkarak's catchment area. (Source: Authors' creation based on [5])

Table 2.2 Land use and surface runoff changing from 1990 to 2020 (Source: Authors' creations based on satellite data and simulations from SWAT model)

No	Land use	1990		2000		2010		2020	
		Ha							
0	No data	1	0	0	0	186	0.2	234	0.2
1	Waterbody	12,033	10.5	12,025	10.5	11,920	10.5	12,057	10.6
2	**Forest**	**23,345**	**20.4**	**27,453**	**24**	**35,274**	**30.9**	**43,286**	**38**
3	Plantation	34,595	30.3	24,112	21.1	21,516	18.9	14,301	12.6
4	Open area	156	0.1	146	0.1	748	0.7	315	0.3
5	Built-up area	1294	1.1	1720	1.5	2266	2	4805	4.2
6	Paddy field	33,026	28.9	26,411	23.1	16,706	14.7	16,909	14.8
7	Shrub	604	0.5	15,527.	13.6	12,272	10.8	9205	8.1
8	Moor	9116	8	6776	5.9	13,282	11.7	13,058	11.5
9	**Surface runoff (mm/year)**	**483.90**	**18.7**	**464.10**	**18.30**	**431.80**	**17.2**	**400.1**	**15.9**

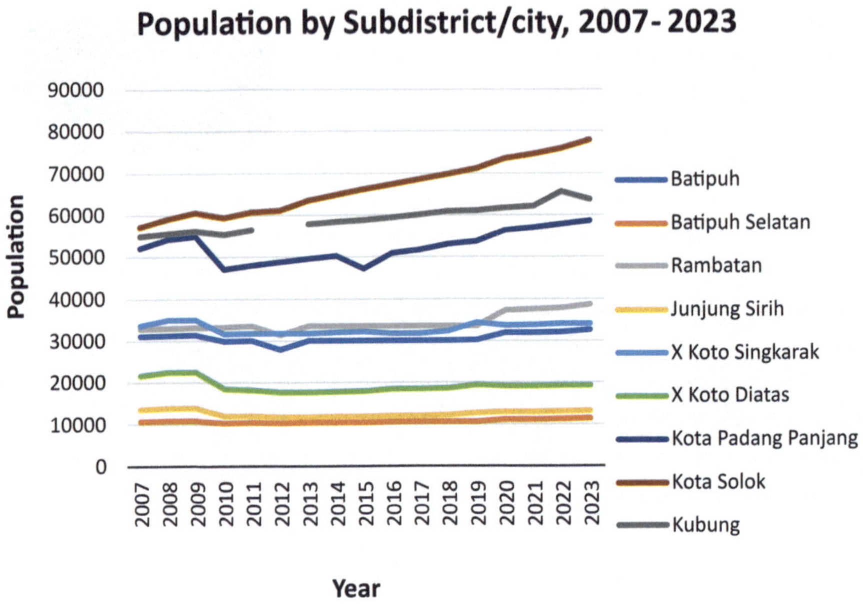

Fig. 2.3 The trends of the increase of the population of cities and *nagaries* in Lake Singkarak's catchment area. (Source: Authors' creation based on [6, 7])

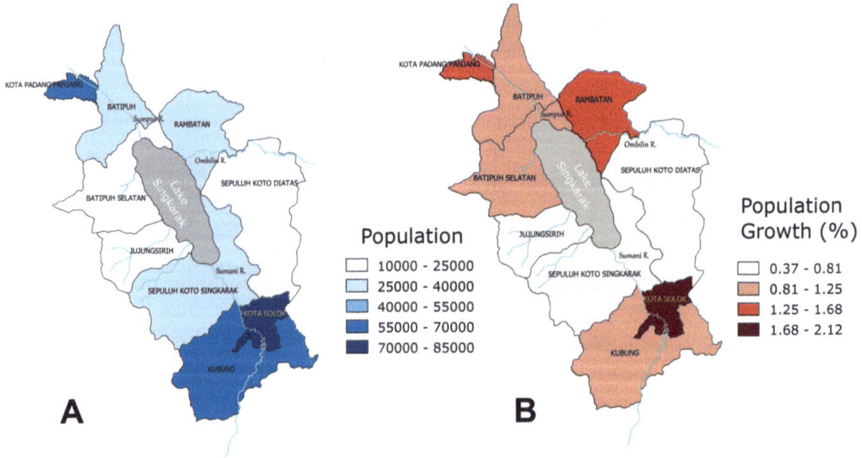

Fig. 2.4 Several demographic parameters of Lake Singkarak's administrative area. (**a**) Population density; (**b**) annual population growth rate. (Source: Authors' creation based on [6, 7])

2.4.2 Resource Use and Anthropogenic Activities in the Lake's Watersheds

Local people mostly use Lake Singkarak for a clean water source, outlet for their waste, capture fisheries, and a local tourism spot which focuses on enjoying the natural scenery and culinary activities [8]. Among the uses, the dominant activity is capture fisheries of *ikan billih* (*Mystacoleucus padangensis*), in which approximately 5000 fishing households are involved [9].

The fisheries stakeholders include lift and non-lift net fishers, investors, and local fisheries' governmental agencies [9]. Most fishers are artisanal fishermen who engage in simple fishing methods using gillnets or traps (ibid). The traps, locally named *alahan*, are used by the fishers to trap the fish in the streams and then fence the stream to capture them. Unfortunately, most fishers combine traps with natural poison, electric fishing, and sometimes application of small amounts of dynamite, making them unsustainable [9]. Further, lift nets with small mesh sizes have also been established, adding severe fishing effort and thus pressure to the fish stocks (Fig. 2.5). A recent survey shows that lift nets reached 317 units in 2022 and more than 400 units in 2024 [10, 11]. The nets, particularly the small mesh-size nets, are considered non-selective fishing gear, which capture the juvenile, destroying its natural recruitment process, with the probability of capture ranging from 94% to 99.8% [12].

Another essential use of the lake is energy generation by a hydropower company. The company moved the natural outlet (*Ombilin* River Mouth) to its intake in *Guguk Malalo Nagari* (Fig. 2.6). Starting their operation in 1998, the company has made Lake Singkarak a semi-natural lake due to its water level management.

There have been severe conflicts of interest between local people and hydropower operations. The locals generally accuse the hydropower operation of causing turbid waters, macroplastic depositions, *ikan bilih* decline, and other environmental problems [8]. However, as with the turbidity topics, none of the claims has been scientifically proven until now (details in the next sections).

Besides fisheries, the lake offers potential tourism services due to its beautiful natural scenery and unique cultural ambiance [13]. However, in general, this

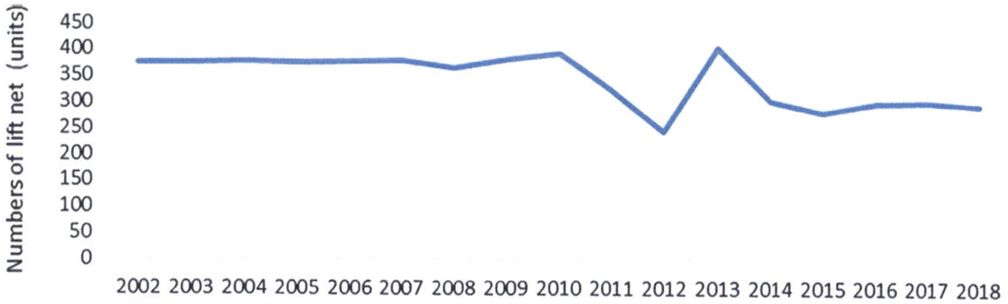

Fig. 2.5 Number of lift nets around Lake Singkarak from 2002 to 2018. (Source: Authors' creation based on [10])

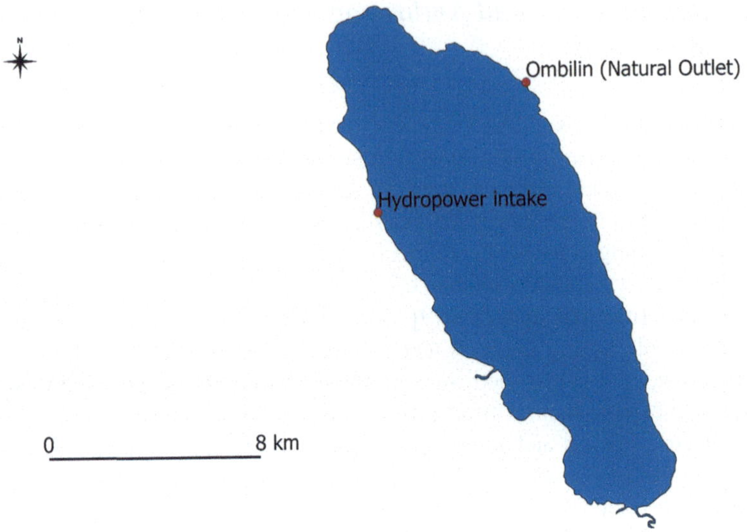

Fig. 2.6 The natural outlet (*Ombilin* River) and the hydropower company's intake. (Source: Authors' creation based on the base map provided by Research Centre for Limnology and Water Resources BRIN, 2025)

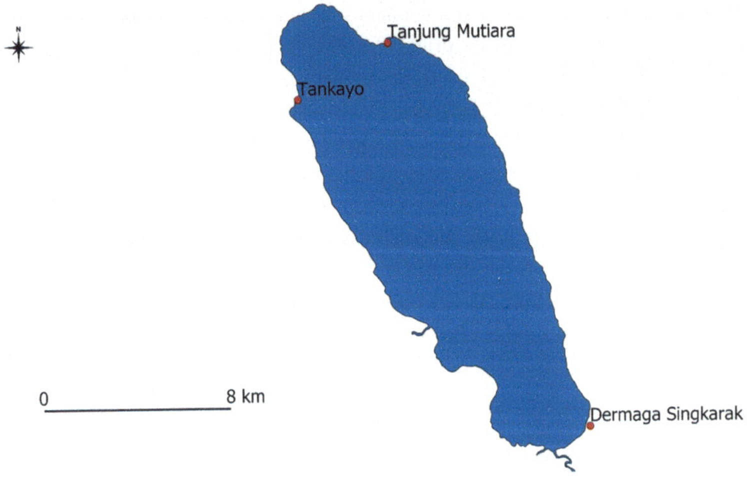

Fig. 2.7 Tourism sites in Lake Singkarak

potential has not been fully developed; in an effort to boost the tourists' visitation rate and local economies, the local government created *Tour de Singkarak*, an international bicycle event [13]. There are a few identified tourist sites in the lake area: Dermaga Singkarak (port), Tanjung Mutiara (a flat beach with boats and water sports activities), and Tankayo (an eco-lodge) (Fig. 2.7).

However, there are very limited publications on the lake's tourism, and secondary data on the lake's tourist visitation is not available. In the statistical records, the number of tourists is mixed with those who visit other tourist sites in the districts, impeding further analysis of the resources used by tourism activities.

We perform a simple SWOT analysis using data obtained from a short primary survey in April 2024 and [14] (Table 2.1). From the analysis, we note that the development of sustainable tourism is hampered by both institutional and infrastructure challenges such as plastic waste management, inadequacy of promotion, and public transportation. The development of ecotourism, cultural tourism, and community-based culinary tourism is essential to enrich the visitor experience. We also found that improving infrastructure accessibility, including transportation networks, alongside strategic digital promotion via social media, official websites, and partnerships with travel agencies, is essential to elevate awareness of the lake as a touristic site.

During the imperialism era (1596–1842), the lake's watershed, particularly the Paninggahan watershed, was transformed into a coffee plantation [15]. After 1958, the area was utilized as secondary forest and then altered into farming and horticulture land such as paddy fields, orchards, clovers, and candlenuts (ibid). As the dominant land use, paddy fields occupied approximately 17% of the watershed area, while horticulture was practiced in 9% of the area, and 12% was utilized for other crops [16]. Further detail of the land use is provided in Chap. 5.

Agricultural activities involve the use of various pesticides, including insecticides, herbicides, fungicides, and molluscicides. Herbicides, fungicides, and molluscicides are generally used only during the growing season of rice, chili, and shallots. Unlike the other three pesticides, insecticides are applied more frequently to control insect pests, especially pyrethroids for shallots, which are added on or once every other day. Pesticides such as permethrin are also used to facilitate the collection of shrimps from the lake.

2.5 Current State

2.5.1 Biophysical Conditions

Based on our research conducted in 2019 and 2022–2024, we can conclude that *ikan bilih's* habitat is still in relatively good condition to support its sustainability. Physical parameters such as pH, turbidity, temperature, and dissolved oxygen (DO) support the fish's well-being.

Ikan bilih lives in clear waters with relatively low temperatures ranging from 26.0 to 28.0 °C and in the littoral area with gravel and sandy substrate [16]. Water temperature has been demonstrated to have a significant impact on fish survival and distribution [17]. Our measurement of the DO values in Lake Singkarak, ranging from 4.56 milligrams per liter (mg/L) to 7.02 mg/L., is within the ambient level to support *ikan bilih* lives. The measurement of pH of the water in the lake's surface area and at the Secchi depth ranges from 8.45 to 9.43. pH level is associated with

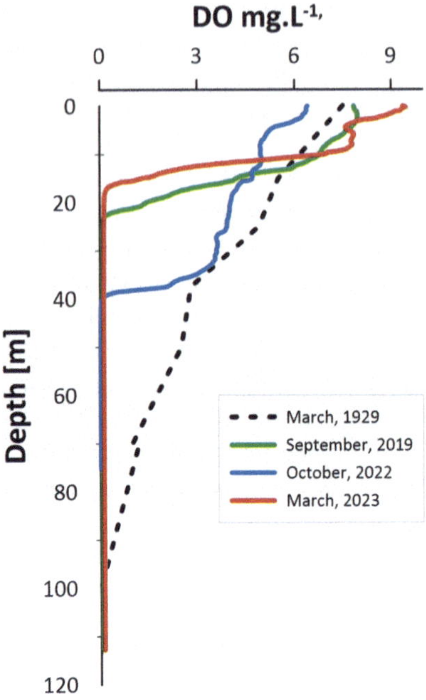

Fig. 2.8 Vertical dissolved oxygen profiles for 1929 [19] and field observation in 2019, 2022, and 2023

the growth performance of fish in a controlled environment, where highly acidic or alkaline waters can inhibit fish growth by disrupting their respiratory functions [18]. However, in a general lake context, considerably higher pH is associated with higher fish species richness [19].

While the surface water of the lake remains healthy for fish habitats, both natural and anthropogenic activities affect oxygen levels in the deeper waters, which is distinctively presented by the shoaling of anoxic water column, indicating the depth of that oxygen-depleted water layer has decreased (Fig. 2.8). In 1929, this boundary was about 100 meters [19] Nevertheless, our observation in 2019 showed that the layer had uplifted to about 22 meters. The fluctuated trends were observed in recent years ranging from about 38 meters in 2022 to 16 meters in 2023. These shifts indicate that the depth boundary between the oxygen-rich and oxygen-poor water layers has become much shallower, raising concerns about the declining water quality conditions in the deeper water layer.

However, although habitat conditions are still relatively favorable for the fish, the stock's decline is definitive. Analysis by [20] revealed that the commonly used mesh sizes are not sustainable as they catch smaller fish than the length of first gonadal maturity. It indicates that overfishing occurs. Further, [21] found that the stock is already in a highly vulnerable state of overfishing, as indicated by the high productivity index, more than its susceptibility index. The overfishing status also resonates in various publications such as [22–24].

At the same time, domestic activities in the catchment area also have obvious effects. The abundance of Coliform bacteria by Tanah Datar Environmental Agency

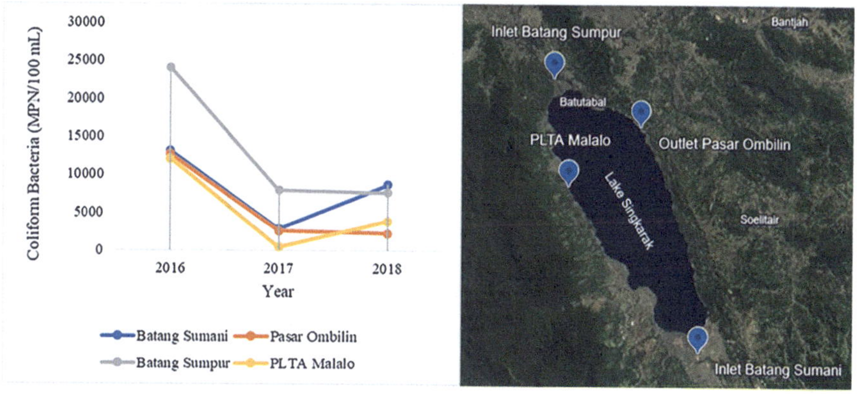

Fig. 2.9 The abundance of Coliform bacteria in several sampling points. (Source: Authors' creation based on [29])

[25] elucidated that there have been indications of high organic contamination as indicated by the high abundance of bacteria, especially in the area surrounding Sumani and Sumpur Rivers, which carries the organic material load from the settlements in Solok and Padang Panjang Cities (Figs. 2.2 and 2.9). The organic material load and bacterial activities can also create turbid waters due to its proliferation characteristic aggregates can develop, which disperse light [26, 27]. Further research is required to confirm this hypothesis.

Further, the analysis revealed that the surface water of the lake contained diazinon residues–a type of pesticide from the organophosphate group, indicating the impacts of agricultural practices [28]. This residue had the highest concentrations at Sumpur area–one of the lake's inlets, reaching 30 μg/L, with an overall average of 13.67 μg/L (Fig. 2.10) [28]. Other types of pesticides, such as permethrin from the pyrethroid group, were also detected at all research stations, except Batu Taba. Again, we also observed the highest concentration of permethrin in the Sumpur area [28].

The residual concentrations of permethrin, cypermethrin, and chlorpyrifos detected in the water of Lake Singkarak exceed the maximum allowable levels set by the Environmental Quality Standards (EQS) [33]. Further, the detection of pesticide residues in the water matrices carries further effects such as sedimental accumulation and bioaccumulation in the bodies of vital organisms such as edible mussels and *ikan bilih* [29, 30]. Considering these impacts, an ecological and health risks assessment of pesticide residues in biotic and abiotic components in the lake is pivotal.

More importantly, considering Lake Singkarak is already prone to organic material enrichment from natural volcanic activities, limiting anthropogenic-based organic and inorganic material loading from agricultural and domestic activities is imperative. Although the lake's current trophic index ranges from mesotrophic–medium trophic level to early eutrophic [31, 32], should business as usual continue, hazards of eutrophication and pesticide pollution are likely inevitable. Thus, we

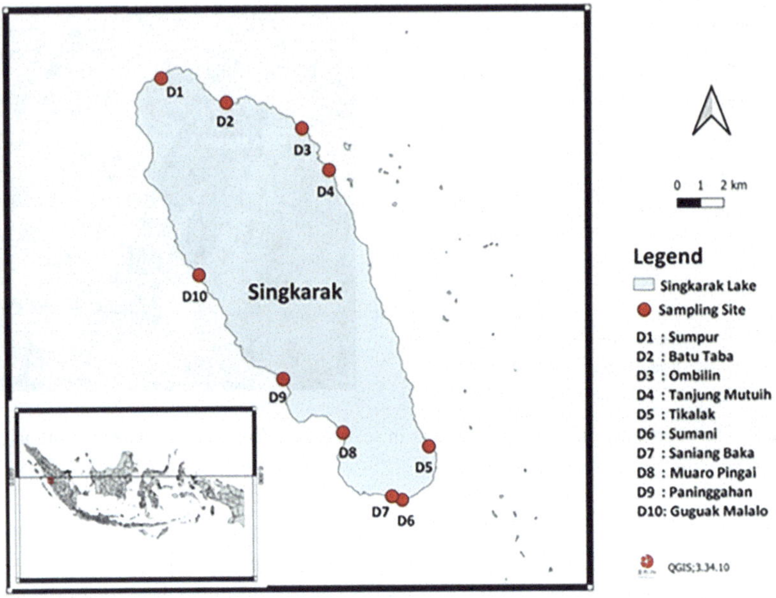

Fig. 2.10 Sampling site for pesticide residues. (Source: authors' creation based on [27])

suggest that routine monitoring aligns with substantial institutional approaches conducted by the authorities.

2.5.2 *Institutional Arrangements*

The institutional arrangements in Lake Singkarak management remain unclear at the time of writing. Although the status of the lake as one of the national priority lakes to undergo restoration [39] has been established, it does clarify the sharing of responsibilities between local and central government at the operational level.

De Jure, with the status as a national priority lake, the central government becomes the principal manager of Lake Singkarak. Management tasks can be transferred and operated by the local governments. Further, considering that the lake is in two districts at the collective level, the provincial government is currently the main management actor. However, there is still confusion about the institutional arrangements on funding for the restoration project and how issued policies are implemented on the operational level. Furthermore, the role of the authority of the *nagari* government in the overall management picture is still unclear.

The institutional analysis of *ikan bilih* fishery is a perfect example of how unclear institutional arrangements affect the biophysical realm. The institutional analysis using the Institutional Analysis and Development (IAD) framework by [9] revealed that the clash between national and local government policies resulted in the dysfunction of lift net limitation management action, which further impedes the establishment of the lake's sustainable capture fishery. Moreover, continuing conflicts between management actors (the hydropower company and fishers) also challenge establishing stakeholders' consensus.

Considering this circumstance, further and more detailed studies on the institutional analysis of lake management, not only on *ikan bilih* fishery, are pivotal. Detailed analysis of how to articulate the issued policies at the operational level in the polycentric governance system of Lake Singkarak is also imperative, as proposed by [33] in a neighboring lake.

2.6 Conclusion

Overall, Lake Singkarak is a case study of wicked management problems in tropical lake systems, which is similar to Lake Maninjau's case study, the neighboring lake [33, 34]. This situation is worsened by the uncertainties generated by climate change and other anthropogenic drivers. To address these challenges, future research on sustainable lake management should be interdisciplinary or even transdisciplinary. Thus, research schemes concerning the lake should be designed to facilitate the interplay between management sectors and stakeholders' interests, and the multisectoral DSS project serves as an example.

References

1. Bemmelen RWV. General geology of Indonesia and adjacent archipelagos. The geology of Indonesia. Batavia: The Hague, Government Printing Office; 1949. https://doi.org/10.1385/0-89603-489-5:219.
2. Zen MT. Structural origin of Lake Singkarak in Central Sumatra. Bull Volcanol. 1971;35:453–61. https://doi.org/10.1007/BF02596966.
3. Wils K, Daryono MR, Praet N, Santoso AB, Dianto A, Schmidt S, et al. The sediments of Lake Singkarak and Lake Maninjau in West Sumatra reveal their earthquake, volcanic and rainfall history. Sediment Geol. 2021;416:105863. https://doi.org/10.1016/j.sedgeo.2021.105863.
4. Setiawan F. Long-term change of water quality in Lake Singkarak observed from remote sensing data. IOP Conf Ser: Earth Environ Sci. Article in press. IOP Publishing. 2024;
5. Peranginangin N, Sakthivadivel R, Scott NR, Kendy E, Steenhuis TS. Water accounting for conjunctive groundwater/surface water management: case of the Singkarak-Ombilin River basin, Indonesia. J Hydrol (Amst). 2004;292:1–22. https://doi.org/10.1016/j.jhydrol.2003.12.018.
6. Pemerintah Daerah Kabupaten Tanah Datar. Kabupaten Tanah Datar dalam Angka. Kabupaten Tanah Datar; 2023.
7. BPS Kota Solok. Kota Solok dalam Angka. BPS Kota Solok; 2023. Available: https://solokkab.bps.go.id/id/publication/2023/02/28/bdcf29eec9b03e0f110ca618/kabupaten-solok-dalam-angka-2023.html
8. Novianti R, Yuniarti I, Ajie GS, Setiawan F, Triharyuni S, Afandi AY. System analysis and research priorities towards sustainable management of Lake Singkarak, Indonesia. In progress. IOP Conf Series: Earth Environ Science. IOP Publishing; 2023.

9. Wicaksono RB. Analisis sistem kelembagaan dan prioritasisasi aksi manajemen perikanan bilih di Danau Singkarak. Unpublished. IPB University. 2024.
10. Agency WSF. Survey jumlah bagan. Unpublished data. Padang; 2024.
11. Fachri F, Hafil M. 317 Unit Bagan di Danau Singkarak Ancam Kelestarian Ikan Bilih. Pengoperasian bagan di Danau Singkarak mengancam ikan bilih. Republika online. 24 Nov 2022. Available: https://news.republika.co.id/berita/rlur0z430/317-unit-bagan-di-danau-singkarak-ancam-kelestarian-ikan-bilih
12. Warsa A, Tjajo DWH, Astuti LP. Upaya Konservasi Ikan Bilih (Mystacoleucus padangensis) dengan Penetapan Ukuran Tangkap dan Suaka di Danau Singkarak. Jurnal Airaha. 2020;9:036–48. https://doi.org/10.15578/ja.v9i01.136.
13. Yunitasari R. Analisis persepsi kondisi pariwisata dan valuasi ekonomi manajelem penanganan sampah makroplastik Danau Singkarak. Unpublished thesis. Unpublished thesis. IPB University; 2024.
14. Pujilestari N, Tarigan SD, Subagyono DK. Analisis Potensi Produksi Air pada Beberapa Skenario Penggunaan Lahan di Daerah Aliran Sungai Paninggahan-Singkarak. Journal Tanah dan Iklim. 2011:13–22. Available: https://repository.pertanian.go.id/server/api/core/bitstreams/842cc4c2-29f0-411a-ad38-113428f2ec34/content
15. Andrea T. Valuing the impact of floods on rice field areas (case study: Paninggahan Village, Junjung Sirih regency). Undergraduate thesis. Universitas Andalas. In progress. Universitas Andalas; 2025.
16. Nasution SH, Suhaemi Syawal M, Akhdiana I. Distribution and habitat characteristics of endemic fish Bilih (Mystacoleucus padangensis Blkr.) in Lake Singkarak, West Sumatra. IOP Conf Ser Earth Environ Sci. 2022;1036:0–14. https://doi.org/10.1088/1755-1315/1036/1/012120.
17. Missaghi S, Hondzo M, Herb W. Prediction of lake water temperature, dissolved oxygen, and fish habitat under changing climate. Clim Chang. 2017;141:747–57. https://doi.org/10.1007/s10584-017-1916-1.
18. Rago PJ, Wiener JG. Does pH affect fish species richness when Lake area is considered? Trans Am Fish Soc. 1986;115:438–47. https://doi.org/10.1577/1548-8659(1986)115<438:dpafsr>2.0.co;2.
19. Rutner F. Hydrographische und hydrochemische Beobachtungen auf Java, Sumatra und Bali. Arch. f. Hydrobiol. 1931;Suppl. 8. Stuttgart: Schweizerbart
20. Triharyuni S, Rahmadi P, Puspasari R, Rachmawati PF, Prianto E. Vulnerability of endemic bilih fish Mystacoleucus padangensis Bleeker, 1852 in Lake Singkarak, West Sumatra, Indonesia. IOP Conf Ser Earth Environ Sci. Institute of Physics. 2022; https://doi.org/10.1088/1755-1315/1119/1/012010.
21. Perret SR, Yuerlita. Adapting to declining fish resources: the differentiation of livelihood systems and fishing strategies in Singkarak Lake's fishing community, West Sumatra. Reg Environ Change. 2014;14:1203–14. https://doi.org/10.1007/s10113-013-0554-z.
22. Warsa A, Krismono. Management of Mystacoleucus padangensis population by the determination of the optimum length at first capture for its exploitation in Lake Singkarak, West Sumatra-Indonesia. IOP Conf Ser Earth Environ Sci. 2020;535 https://doi.org/10.1088/1755-1315/535/1/012044.
23. Warsa A, Hendrawan ALS, Krismono. Penetapan ukuran mata jarin langli untuk penangkapan bilih (Mystacoleucus padangensis) di Danau Singkarak. J Fish Mar Res. 2020;4:178–86. https://doi.org/10.21776/ub.jfmr.2020.004.01.25.
24. Warsa A, Astuti DWHTLP. Upaya Konservasi Ikan Bilih (Mystacoleucus padangensis) dengan Penetapan Ukuran Tangkap dan Suaka di Danau Singkarak. Airaha. 2020;IX:036–48.
25. Pemerintah Provinsi Sumatra Barat. Dokumen Informasi Kinerja Pengelolaan Lingkungan Hidup Daerah Provinsi Sumatra Barat. Padang; 2019.
26. Azhdarpoor A, Salehi N, Heidari H, Sarmadipour M, Mahmoudian H. Relationship between turbidity and microbial load of water in Salman Farsi Dam Reservoir. J Environ Pollut Manag.

2019;2:2–5. Available: https://article.scholarena.com/Relationship-between-Turbidity-and-Microbial-Load-of-Water-in-Salman-Farsi-Dam-Reservoir.pdf
27. Utami RR, Aiman Ibrahim, Nugraha Ardiwinata A, Suhaemi Syawal M, Mardiati Y, Suryono T, et al. Assessing pesticide contamination in Lake Singkarak: ecological risk prioritization and agricultural land use connections. J Great Lakes Res. In progress. 2025;
28. Lópe J, Halleux V, Bentele H, Jorge Buxadé Villalba PfE, Fiocchi P, Wiezik M, et al. Proposal for a Directive of the European Parliament and of the Council Amending Directive 2000/60/EC Establishing a Framework for Community Action in the field of Water Policy, Directive 2006/118/EC on the Protection of Groundwater Against Pollution and Deterioration and Directive 2008/105/EC on Environmental Quality Standards in the Field of Water Policy. BRIEFING EU Legislation in Progress. 2022.
29. Syawal MS, Ibrahim A, Yustiawati, Nasution SH, Taufik I, Saraswati M, et al. Organophosphate pesticide residues in surface water and bilih fish (*Mystacoleucus padangensis* Blkr.) in Lake Singkarak, West Sumatra. IOP Conf Ser: Earth Environ Sci. Institute of Physics. 2023; https://doi.org/10.1088/1755-1315/1221/1/012080.
30. Ibrahim A, Suhaemi Syawal M, Nugraha Ardiwinata A, Sugiarti, Awaludin Adam M, Fitrada W, et al. Organochlorine and pyrethroid residue in fish and sediment of Lake Singkarak, a tropical deep lake. LIMNOTEK Perairan Darat Tropis di Indonesia. 2023; https://doi.org/10.55981/limnotek.2023.2084.
31. Setiawan F, Jiang D, Hamzah R, Matsushita B. Inland water trophic state identification using remote sensing data and machine learning approach. IOP Conf Ser Earth Environ Sci. 2022;1062 https://doi.org/10.1088/1755-1315/1062/1/012033.
32. Lukman L, Syawal MS, Maghfiroh M. Sumatran major lakes: Limnological overviews. IOP Conf Ser Earth Environ Sci. 2020; https://doi.org/10.1088/1755-1315/535/1/012064.
33. Yuniarti I, Barnes C, Glenk K, Sutrisno. Challenges for the development of environmentally sustainable cage culture farming in Lake Maninjau, Indonesia: an institutional perspective. Ecosyst People. 2021;17:248–63. https://doi.org/10.1080/26395916.2021.1921848.
34. Yuniarti I, Barnes C, Glenk K, McVittie A. Towards sustainable lake ecosystem-based management– lessons learned from interdisciplinary research of cage aquaculture management in Lake Maninjau. In: Triyanti A, Djalante R, Marfai MA, Indrawan M, Nurhidayah L, editors. Environmental governance in Indonesia. Springer; 2023.

Chapter 3
The Analytical Framework for Participatory Problem Prioritizations: A Fundamental of the Decision Support System (DSS)

Guruh Satria Ajie, Ivana Yuniarti, and Yuerlita

Abstract This chapter elucidates how the foundation framework of the decision support system (DSS) was built using an analytical framework, referring to the conceptual model of Ecosystem-Based Management-Drivers-Pressures-States-Ecosystem Services-Responses (EBM-DPSER). It starts with the introduction of the theoretical foundation of the framework and runs through the quantification of each of the framework's component. The framework provides robust scientific fundamentals and legitimacy for the DSS since it was constructed using stakeholder and expert-based knowledge. It portrays the dynamics of lake ecosystems and the complex interactions of their constituent components (both abiotic and biotic components and the socio-economic systems). The quantification also results in the weights of each management issue and its components, highlighting their relative importance. Besides becoming the essential inputs in the DSS, the results also present the management problem prioritization, guiding the selection of essential management responses.

Keywords EBM-DPSER framework · DPSIR framework · Lake · Lake management · Decision support system · Ecosystem

---

G. S. Ajie (✉)
Research Centre for Limnology and Water Resources, National Research and Innovation Agency (BRIN), Cibinong, Indonesia
e-mail: guruh.satria.ajie@brin.go.id

I. Yuniarti
Research Centre for Environmental Archaeology, Maritime Archaeology, and Cultural Sustainability, National Research and Innovation Agency (BRIN), Cibinong, Indonesia

Yuerlita
Department of Agribusiness, Faculty of Socio-economic of Agriculture, Universitas Andalas, Padang, Indonesia

K. Glenk et al. (eds.), *Decision Support Systems for Multisectoral Tropical Lake Management in Indonesia*, https://doi.org/10.1007/978-3-032-11268-2_3

3.1 Chapter Introduction

This chapter elucidates how the foundation framework of the DSS was built. We constructed an analytical framework referring to the conceptual model of Ecosystem-Based Management-Drivers-Pressures-States-Ecosystem Services-Responses (EBM-DPSER) [1]. The ecosystem services section of the framework was selected because it can accommodate quantitative approaches aligned with the DSS. The framework is a modification of the popular Drivers-Pressures-States-Impacts-Responses (DPSIR) framework developed by the European Environmental Agency (EEA).

Our framework provides robust scientific fundamentals and legitimacy for the DSS since it was constructed using stakeholder and expert-based knowledge. It portrays the dynamics of lake ecosystems and the complex interactions of their constituent components (both abiotic and biotic components and the socio-economic systems [2]). More importantly, the flexibility of the framework to adapt to stakeholders' opinions and knowledge guided us to prioritize important models to be built to provide evidence-based support in the face of stakeholders' perceptions of lake challenges [3]. While considering stakeholder interests, we took precautionary steps to avoid being led by the stakeholders' interests by conducting expert panel consultations and by referring to the observational field data. Our framework is also fundamental to calculating a lake management sustainability index (SI), which provides a principal monitoring indicator (details in Chap. 8).

The whole system analysis to construct the framework is called participatory environmental system analysis (ESA). It was developed initially in the United States of America (USA) with the beginning of rational development planning theory in the 1950s, which encouraged the development of problem-solving-oriented policy-making using systematic, scientific, and interdisciplinary approaches. The ESA is an integrative, interdisciplinary, and participatory approach using various methods to analyze the environment based on causes, mechanisms, impacts, and solutions that can be offered to improve current conditions [4].

3.2 Introduction to Participatory Framework DPSIR

The basic participatory DPSIR framework (Fig. 3.1) was constructed following [6–8] with an important modification using stakeholders' involvement. As a brief introduction, the framework consists of several components: drivers, pressures, states, impacts, and responses. Drivers are indirect causes of the state of the ecosystem resulting from the needs of society due to socio-economic development (e.g., population growth). Pressures are human activities that directly impact lake ecosystem state changes, such as overfishing, illegal fishing methods, pollutant input loads, and erosion from the lake water catchment area (WCA). States are defined as the current condition of the lake ecosystem (i.e., water quality). Meanwhile, impacts are

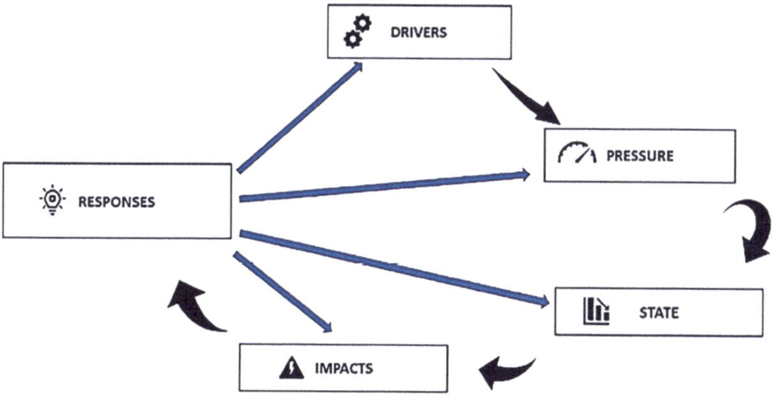

Fig. 3.1 Schematic of the framework DPSIR. (Source: Authors' creation based on the DPSIR framework by [5])

the environmental services and functions derived from the lake ecosystem for human well-being. In this book, we replace the impacts with the ecosystem services (ES) component to directly link it with the ES framework. Lastly, responses are lake management actions that target other framework components to produce a desirable impact for the communities. We removed the response part for the DSS construction as the response part will be the DSS's output.

3.3 Methods

The general methods illustrated in Fig. 3.2 present the inclusion of stakeholders' perspectives, local knowledge, expert knowledge, and scientific knowledge to construct the framework. The framework is then used to construct scenario-based modelling as the input to the DSS.

3.3.1 Framework Construction: Stakeholders' Involvement and Expert Panel Consultation

We began the research by conducting structured interviews directly with local stakeholders (i.e., Tanah Datar and Solok districts' planning agencies, environmental agencies, fisheries agencies, tourism agencies, fishers, the chief of Lake Singkarak's Management Team, and *wali nagaries*– customary village leaders). In the interviews, we explored both stakeholders' perspectives on the lake management issues and collected information on their prioritization of issues and actions.

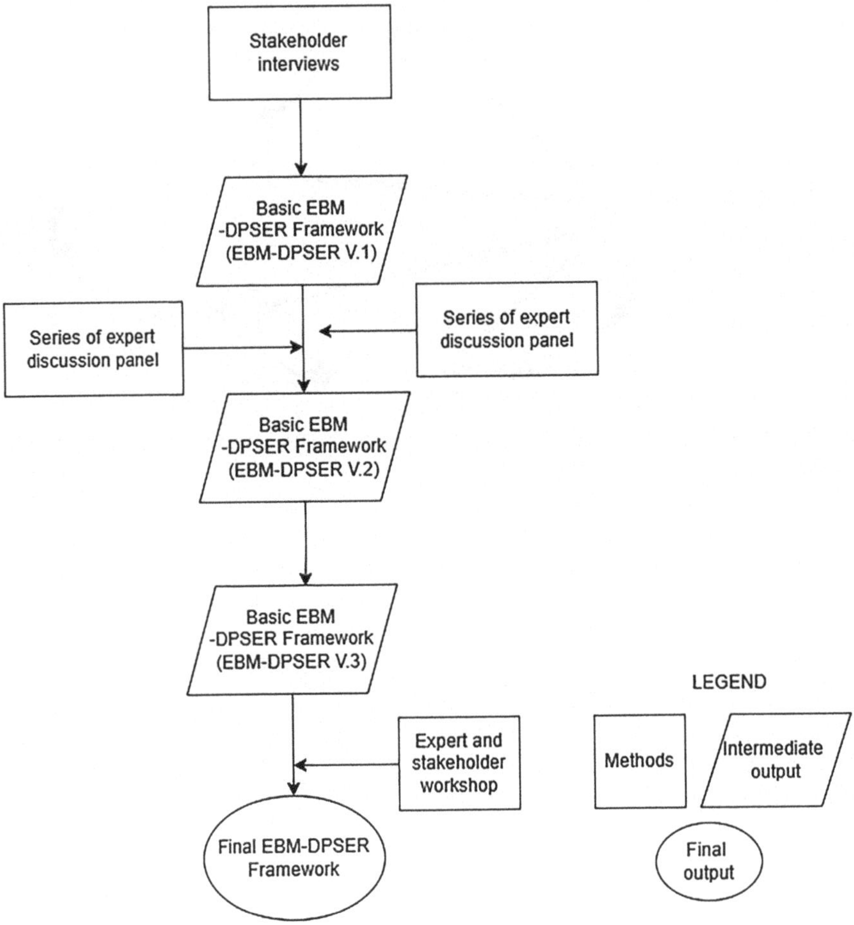

Fig. 3.2 General methods to construct the EBM-DPSER framework and the DSS

The interview data helped to sketch an initial framework, referred to as EBM-DPSER V.1.

To follow up on the process, we arranged a series of expert panel consultations to evaluate the drafted initial framework. Firstly, we distributed 135 online questionnaires to the selected experts chosen to reflect a range of expertise, evidenced through appropriate education and documented works concerning the lake (Fig. 3.3). We received 25 completed questionnaires. The questions asked about the experts themselves, included Likert-scaled expert-based perception question on the current conditions of the lake and on management issues prioritization; and concerned information on the current state of the lake, including available data.

Secondly, using the information we received from the experts and some of our initial models (e.g., habitat models), we modified the initial framework to create a

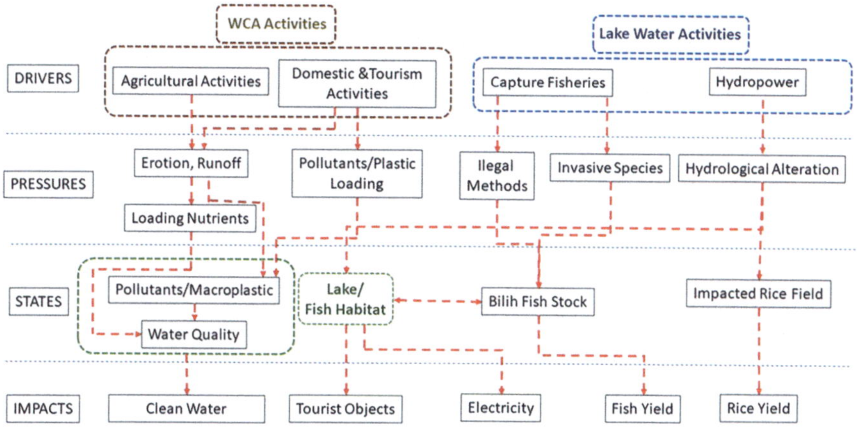

Fig. 3.3 The final EBM-DPSI framework

new version referred to as DPSIR V.2. This version was presented in an online workshop attended by 25 experts. We also presented the initial framework to make the attendees understand the process better. The discussion was aimed at seeking common agreement. The result and agreed framework from the workshop, named DPSIR V.3, was then presented and discussed with a larger stakeholders group in an offline workshop. The result of the workshop was then used as the final framework.

This stakeholder workshop was held in Tanah Datar's Planning Agency in September 2023. Its aim was to generate a common understanding among the stakeholders of the framework and the management issues. We were able to establish an agreed-upon framework, although management issues and their reflection remained unresolved, such as the effects of hydropower companies on macroplastic depositions. The unresolved issues highlighted the importance of building technical models and the use of the DSS to bridge the disputes.

3.3.2 *Problem Prioritizations*

The weighting of the lake management problems in the framework was done using the analytic hierarchy process (AHP) method [6, 7]. This method combined qualitative and quantitative assessments to obtain a representative and definitive final weight score.

The first step to conducting such an analysis is to create a hierarchical or network structure from the problem statement. In the hierarchy, there are some parts to be involved: main objectives, criteria, sub-criteria, and alternatives. Next, pairwise comparisons were made to form relationships in the analyzed structure (Table 3.1). The numerical values used for pairwise comparison are obtained from a comparison scale of 1 to 9, with the meaning of each number listed in Table 3.2. The overall

Table 3.1 Pairwise comparison matrix example

C	**A_1**	**A_2**	**……**	**An**
A_1	a_{11}	a_{12}	…..	a_{1n}
A_2	a_{21}	a_{22}	…..	a_{2n}
…..	…..	…..	…..	…..
An	a_{n1}	a_{n2}	…..	a_{nn}

Table 3.2 Pair comparison scale on the degree of importance

Degree of importance	Definition	Explanation
1	Equally important	Both choices have the same importance
3	A little more important	Assessors slightly favor one of the choices
5	More important	Assessors are more inclined to favor one of the choices
7	Very important	Assessors are very inclined to the one of choices with real dominance
9	Absolutely more important	Assessors' absolute choice is preferred, with the highest level of confidence
2,4,6,8	In-between values	The assessors express a compromise between opinions on one level above and below

result is in the form of the weighting of each criterion after calculating the pairwise comparison matrix and the normalized comparison matrix. In this method, a consistency test for the choices made was done using the consistency ratio (CR), where the calculated value must be less than 0.1 to pass the test. To conduct the test, we first calculated the weight of each main lake management problem, and then we computed the weight of each root of the problem by dividing the results by the number of the roots.

3.4 The Final EBM-DPSER Framework

3.4.1 The Driver Components

The resulting DPSER framework (Fig. 3.3) includes two types of primary socio-economic activity drivers: anthropogenic activities in the water catchment area (WCA) and lake water use related to hydropower generation. Hydropower activities are considered as one of the principal drivers as the stakeholders perceived that its operation has altered hydrological regime of the lake creating further pressures such as macroplastic deposition.

WCA activities consist of three components: agricultural, domestic, and tourism activities. As previously explained, the WCA is located in Solok and Tanah Datar Districts. In Solok District, the anthropogenic influences are dominated by the agricultural sector [8]. In this district, rice cultivation dominates (20.561 ha, producing 182,203 tons of rice in 2023), followed by horticulture (shallot (*Allium cepa* L. var. *aggregatum*) and keriting pepper (*Capsicum annuum* var. *acuminatum*), with total production of about 216, 418, and 37.237 tons, respectively, in the same year (ibid). In the horticulture sector, coffee is the main crop occupying an area of 8759 ha.

To a much lesser extent, Solok District also affects the lake via tourism and domestic influences. The district has 299 tourist attractions consisting of 127 natural and 172 other cultural attractions. There were 49 tourist accommodations, with as many as 1.1 million tourists registered to visit in 2023 (ibid).

In the Tanah Datar district area, agriculture, domestic, and tourism activities are the principal human influences on the WCA. The agricultural sector is dominated by rice cultivation at about 53,055.59 ha in 2022 [9]. The district had approximately 195 tourist attractions, 46 accommodations, and 13 tourist companies. However, specific data related to the WCA and the lake are not specifically described [9]. The region has a low population growth rate of only 0.69% annually and a workforce of over 174,000 people, with a total unemployment of 10,000 people.

Capture fisheries, as one of the principal water-based activities, are conducted on tilapia (*Oreochromis niloticus*) and carp (*Cyprinus carpio*) with a total production of 4,756,900 tons and 3,315,400 tons, respectively [10]. The sector employs 859 full-time and 2260 part-time fishermen (ibid). Meanwhile, the only recorded capture of fish species for the Solok District was the capture of 4645 tons of tilapia [8]. These two species are considered non-native species, which can adversely affect the native fish population; therefore, the monitoring of the native population is vital. However, no reliable sources could be obtained about related efforts.

The second activity driver is the operation of the hydropower plant. Built in 1992, a 19-kilometer-long Singkarak hydropower tunnel was created through Mount Merapi to flow water from the lake to the turbine located in Lubuk Alung in the West of Singkarak. To support the operation of the tunnel, the company regulated the lake outlet by constructing Ombilin Weirs. This alteration has been accused of creating frequent floods in the surrounding marginal rice fields. Furthermore, there are strong stakeholder perceptions that this change also negatively affected *ikan bilih* habitat. However, this argument was not scientifically supported by a MaxEnt habitat model simulation and by recent observation of the lake's water quality (see Chap. 6). Thus, we removed this cause–effect relationship in our DSS.

3.4.2 The Pressure and State Components

The main pressure components consist of erosion, runoff, and sedimentation. The results from the satellite image analysis in 2000, 2006, and 2011 showed that there had been a decrease in the primary and secondary forest areas in sub-districts Sumpur and Paninggahan, two of the main villages producing the endemic fish—*ikan bilih* (*Mystacoleucus padangensis*) [11, 12]. In contrast, dry land farming tends to increase in those areas (ibid). However, overall, there was a substantial increase in forest coverage in the watershed area after 2011, as described in Chap. 4.

The improvement has reduced and caused less erosion in the lake catchment area at a rate lower than, with a figure of 10.2 tons/ha/year in those sub-districts (ibid). Further, it is observed that the sedimentation rate of the two main lake inlets (Sumpur and Sumani) reached 30–40 cm/year in 2019 [13]. Meanwhile, it is essential to note that, until the time of writing this chapter, we cannot obtain reliable information on the loading of macro- and microplastics, antibiotics, and other hazardous substances into the lake.

Aligning with the low erosion rate, long-term monitoring satellite data revealed that there is a tendency for improvement of the lake's water clarity, which is highly likely caused by the flushing effect of the operational of Lake Singkarak Hydropower Plant [14]. In addition to this evidence, a recent water quality observation showed that the Secchi depth ranged from 0.25 to 15.4 m and the level of Total Dissolved Solids (TDS) was 0,075 to 0,161 mg/L [15].

Further, the observational data revealed that the lake is categorized as being in oligo and mesotrophic states, with the concentration of total phosphorus (TP) ranging between 0.01 and 0.16 mg/L in early 2000, but rising to the range of 0.08 and 0.64 mg/L over the past eight years. Meanwhile, a recent measurement of total nitrogen (TN) concentrations obtained a very low concentration in the hypolimnion, to a maximum of 1 mg/L on the lake surface, and the maximum chlorophyll-a in this lake can reach 20.91 mg/m^3 (ibid). Thus, in general, the lake water condition was in a decent state.

At the same time, the most recent study on macroplastic deposition presented that the density of the pollutant in the lake bedside ranged from 0.5 to 3 kg/m^2. Nevertheless, there have been no further studies on the deposition pattern of the plastic, although it has become a public and tourist concern [16].

With a focus on the capture fisheries sector, a study about the vulnerability of *ikan bilih* revealed that the greatest pressure on the fish population is overfishing and the use of unsustainable fishing gear [17]. Their maximum sustainability yield model (MSY) calculated that the optimal harvest of *ikan bilih* is at 1126 tons, equivalent to 3753 net units. The most common fishing gears are lift nets (*bagan*) (Fig. 3.4) and *alahan*−traditional traps blocking the small stream areas to create enclosures for the locally migratory fish species (Fig. 3.5), gill nets, and throw nets (ibid). In total, there were 736 pieces of gear in 2015 (340 gill nets, 363 throw nets, and 33 units of Alahan) [18]. More importantly, 534 units of bagan were recorded in 2019 [19]. Bagan itself is considered an unsustainable fishing gear that was once listed as illegal fishing gear for use in Lake Singkarak. It is generally considered unsustainable due to the use of unselective small nets (<1 inch) that have been prohibited for large-scale catches.

Fig. 3.4 Bagan-lift net fishing gear (Source: Authors' personal documentation)

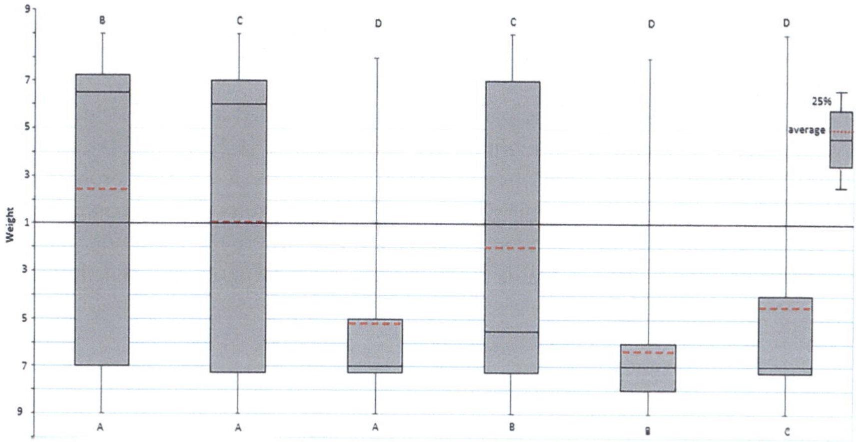

Fig. 3.5 Stakeholders and expert perspectives on lake problems weighting

3.5 The Stakeholder Version of EBM-DPSER Framework (First Version)

Moderately different from the final framework shown in Fig. 3.3, the first version contains four main problems with five roots. The first problem is the decrease in the catch of an endemic and important economic fish species, *ikan bilih* (*Mystacoleucus padangensis*) (coded with **A**), which includes three path roots: unsustainable fishing methods of catching (**A.1**), habitat degradation from microplastic accumulation in the lake area (**A.2**), and invasive fish species to (**A.3**). The second problem is related to the declining water quality (**B**) due to the contaminants or pollutants loading into the lake mainly from its watershed (**B.1**). Some respondents stated that the possible cause of this problem was hydropower operations that caused the accumulation of macroplastic and pollutants at the bottom of the lake (**B.2**). As for the third problem, the garbage-related problem (**C**) is caused by domestic activities that affect lake tourism and the supply of clean water from the lake. The fourth problem concerns the perception of flooding of the marginal rice fields due to the water level regulation by the hydropower company (**D**).

3.6 Expert and Stakeholders' Prioritizations Scale of the Lake Management Problems

Using the results of the direct interviews and the online questionnaires, we then create the expert and stakeholders' prioritization scale for the lake management problems. The weighing result of the lake's problems and its root (Figs. 3.5 and 3.6) illustrated that there was wide heterogeneity among the participants opinions about the relative importance of each problem. However, in general, it can be concluded that solving problems B (water quality), A (*ikan bilih*), and C (macroplastic) were more important than D.

We pointed out that the participants perceived Problem B as the most important problem to solve. Its relative weight to problems A and C is 2.54 and 2.04, respectively. Meanwhile, the weighted result from the comparison between problems C and A yields an equally important value of 1.17. These results slightly contradicted the recent state observation previously defined above, where the *ikan bilih* population is declining, and water quality is moderately fine. Experts' and stakeholders'

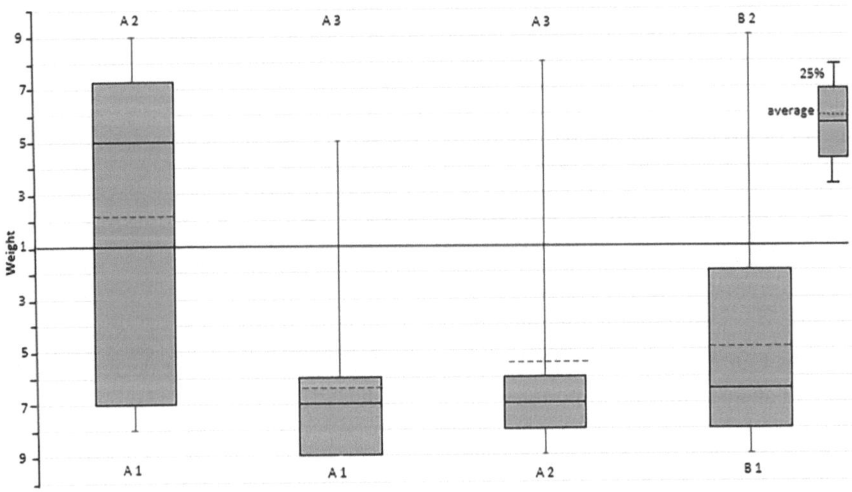

Fig. 3.6 Boxplot on stakeholders' perspective on lake roots of the weighting of the problem

backgrounds and limited access to the most recent and long-term data steered their opinion. The bias created by this matter was mitigated by conducting a short briefing on the current observational states. Then, we created an online workshop to observe the changes in stakeholders' and experts' opinions and incorporated the results of the workshop in the final output.

The participants consider the solutions to ikan bilih (A) root problems A.1 (unsustainable fishing methods) and A.2 (habitat degradation) to be more important than the solutions to A.3 (invasive species), with weights of 6.12 and 4.12, respectively. However, when we compared the importance of A.1 and A.2 solutions, A.1 was perceived to be relatively weaker than A.2, with a weighing value of 2.04. In the water quality problem (B), the respondents regarded the solution to root problem B.1 (watershed source) as more important than B.2 (hydropower-related cause), with a weight value of 6.92.

Next, we performed AHP on the data to weigh the value of each lake management problem and its roots. The pairwise comparison matrix of the main problem (Table 3.3) was filled using the rounding of the average weighting value of the stakeholders. This matrix's last row is the sum of the values from the column above to obtain the divisor number for calculating the normalized pairwise comparison matrix.

The normalized calculation results of this matrix in Table 3.4 show that problem B (water quality) is the main priority of stakeholders, with a weight of almost 45%. Problems A (*ikan bilih*) and C (macroplastic) are the following, with equal weights of 25% and 24%, respectively, trailed by Problem D (flooded marginal rice field) which is their least concern. The small weight of problem D also indicates that, this problem is not the stakeholders' main concern.

Table 3.3 Lake problems pairwise comparison matrix

	A	B	C	D
A	1.00	0.50	1.00	5.00
B	2.00	1.00	2.00	6.00
C	1.00	0.50	1.00	4.00
D	0.20	0.17	0.25	1.00
Sum	4.2	2.2	4.3	16.0

Table 3.4 Lake problems normalized pairwise comparison matrix

	A	B	C	D	Weights	Problem weights
A	0.238	0.231	0.235	0.313	1.017	0.254
B	0.476	0.462	0.471	0.375	1.783	0.446
C	0.238	0.231	0.235	0.250	0.954	0.239
D	0.048	0.077	0.059	0.063	0.246	0.061
Sum					4.000	1.000

The consistency matrix presenting the consistency of the stakeholders' choices (Table 3.5) shows a CR value of 0.008 (<0.1). It indicates that the stakeholder choices in weighing the four main LS management problems were consistent.

The weighing calculation of the underpinning problems of A and B was done in a similar AHP sequence. The results in Tables 3.6 and 3.7 show A.2 (habitat degradation) as the main problem root of A (*ikan bilih*) with a 55% weighting. The underpinning causes A.1 (unsustainable fishing) and A.3 (invasive species), which weighed only 37% and 8%, respectively. The calculation of the consistency matrix in Table 3.8 resulted in a CR value of 0.075 (<0.1). Therefore, the stakeholders' choices regarding the roots of problem A were consistent. Again, it is a surprising result as the current research mentioned that overfishing is the principal cause of the decline of *ikan bilih* catch. We relate this issue to the low response of fisheries experts and the dominance of water quality-related experts in returning the questionnaires. This bias should be overcome with better expert panel selections in future research.

The roots of the problem B (water quality) calculation using the AHP method, shown in Tables 3.9 and 3.10, resulted in B.1 (watershed loading) as the main root of B (water quality) with an 80% weighting. The root problem B.2 (Hydropower cause) weighed only 20%.

Table 3.5 Lake problems consistency calculation matrix

	A	B	C	D	Sum	Weights	S/W	
A	1.017	0.892	0.954	1.229	4.092	1.017	4.025	
B	2.033	1.783	1.908	1.475	7.200	1.783	4.038	
C	1.017	0.892	0.954	0.983	3.846	0.954	4.031	
D	0.203	0.297	0.239	0.246	0.985	0.246	4.006	
							4.025	λmax
							0.008	CI
							0.009	**CR**

Table 3.6 Roots of the lake problems: a pairwise comparison matrix

	A.1	A.2	A.3
A.1	1.00	0.50	6.00
A.2	2.00	1.00	5.00
A.3	0.17	0.20	1.00
Sum	3.17	1.70	12.00

Table 3.7 Roots of the lake problems: a normalized pairwise comparison matrix

	A.1	A.2	A.3	Weights	Norm	Problem root weight
A.1	0.32	0.29	0.50	1.11	0.370	0.094
A.2	0.63	0.59	0.42	1.64	0.545	0.139
A.3	0.05	0.12	0.08	0.25	0.085	0.021
Sum				3.00	1.000	0.254

Table 3.8 Roots of the lake problems: A Lake problems consistency calculation matrix

	A.1	A.2	A.3	Sum	Weights	S/W	
A.1	1.11	0.82	1.52	3.45	1.11	3.108	
A.2	2.22	1.64	1.27	5.12	1.64	3.131	
A.3	0.18	0.33	0.25	0.77	0.25	3.020	
						3.086	λmax
						0.043	CI
						0.075	C/R

Table 3.9 Roots of the lake problems B pairwise comparison matrix

	B.1	B.2
B.1	1.00	4.00
B.2	0.25	1.00
Sum	1.25	5.00

Table 3.10 Roots of lake Singkarak's problems B normalized pairwise comparison matrix

	B.1	B.2	Weights	Norm	Problem root weights
B.1	0.80	0.80	1.60	0.800	0.357
B.2	0.20	0.20	0.40	0.200	0.089
Sum			2.00	1.000	0.446

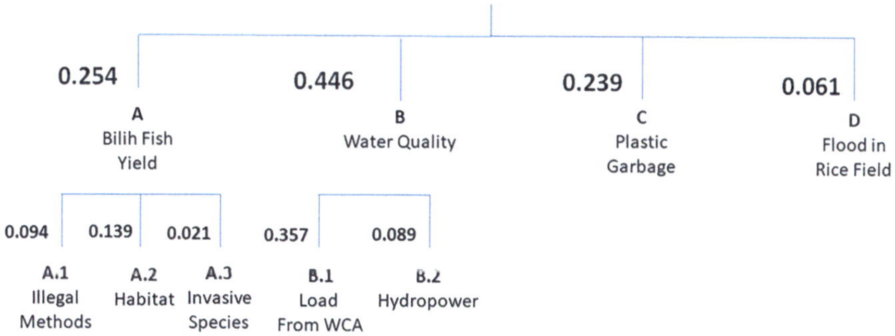

Fig. 3.7 Hierarchical graph of the main problems and their roots

A summary of the lake's main problems and their roots in a hierarchical graph in Fig. 3.7 shows that stakeholders place almost half the weight on the solution to problem B (water quality). Next, for the second and third problems, there is a weighting of only half of problem B (water quality) for problems A (*ikan bilih*) and C (macroplastic). Problem D (flooded marginal rice field) received a relatively small weight. Stakeholders assigned that the highest priority for solution should be given to problem B, with the root of problem B.1 (watershed loading) having a weight value of 0.357. For problem A (*ikan bilih*), the stakeholders argue that A.2 (habitat degradation) is the root problem with a weight of 0.139. Despite the previously discussed biases and limitations, these results are considerably valid to be used as a basis for calculating the Sustainability Index (SI) in the DSS.

3.7 Conclusion

Overall, we have analyzed that water quality and the decline of the ikan bilih population are the principal problems, followed by macroplastic deposition problem. We have also identified the roots of the problems of those matters. Our findings were then used to identify the management action scenarios presented in the DSS, which became the focus of the calculation of benefit-cost analysis and sustainability index (details in Chap. 8). In conclusion, we have pointed out that local knowledge can be integrated with scientific knowledge to provide a robust fundamental framework to create further advanced research, such as the DSS. This illustration contributes significantly to the heated discussion between local and scientific knowledge. Further details related to this issue and discussions on the integration between local and scientific knowledge can be read in Chap. 10.

References

1. Kelble CR, Loomis DK, Lovelace S, Nuttle WK, Ortner PB, Fletcher P, et al. The EBM-DPSER conceptual model: integrating ecosystem services into the DPSIR framework. PLoS One. 2013;8:e70766. https://doi.org/10.1371/journal.pone.0070766.
2. Dubey D, Dutta V. Nutrient enrichment in lake ecosystem and its effects on algae and macrophytes. In: Environmental concerns and sustainable development: Volume 2: Biodiversity, soil and waste management; 2019. https://doi.org/10.1007/978-981-13-6358-0_5.
3. Ovienmhada U, Mouftaou F, Wood D. Inclusive design of earth observation decision support systems for environmental governance: a case study of Lake Nokoué. Front Clim. 2021;3:1–18. https://doi.org/10.3389/fclim.2021.717418.
4. Brown CM, Lund JR, Cai X, Reed PM, Zagona EA, Ostfeld A, et al. The future of water resources systems analysis: toward a scientific framework for sustainable water management. Water Resour Res. 2015:6110–24. https://doi.org/10.1002/2015WR017114.
5. Kristensen P. The DPSIR framework. A comprehensive/detailed assessment of the vulnerability of water resources to environmental change in Africa using the river basin approach; 2004. p. 1–10. Available: http://enviro.lclark.edu:8002/rid=1145949501662_742777852_522/DPSIROverview.pdf
6. Khan AU, Khan AU, Ali Y. Analytical Hierarchy Process (Ahp) and analytic network process methods and their applications: a twenty year review from 2000–2019. Int J Anal Hierarchy Process. 2020;12:369–402. https://doi.org/10.13033/IJAHP.V12I3.822.
7. Veisi H, Deihimfard R, Shahmohammadi A, Hydarzadeh Y. Application of the analytic hierarchy process (AHP) in a multi-criteria selection of agricultural irrigation systems. Agric Water Manage. 2022;267 https://doi.org/10.1016/j.agwat.2022.107619.
8. BPS Kota Solok. Kota Solok dalam Angka. BPS Kota Solok; 2023. Available: https://solokkab.bps.go.id/id/publication/2023/02/28/bdcf29eec9b03e0f110ca618/kabupaten-solok-dalam-angka-2023.html
9. Kabupaten Tanah Datar. Kabupaten Tanah Datar dalam Angka. Kabupaten Tanah Datar; 2023.

10. Dinas Perikanan Kabupaten Tanah Datar. Data Perikanan Kabupaten Tanah Datar 2023. unpublished. 2023.
11. Azmi R. Kajian tataguna lahan untuk memperkecil erosi di sub das paninggahan menggunakan metode usle dan kaitannya dengan keberlangsungan sumber daya air Danau Singkarak. Unpublished thesis. Universitas Bung Hatta; 2014.
12. Fasiven G. Rekayasa tataguna lahan untuk memperkecil erosi dengan metode USLE di Sub DAS Sumpur dalam rangka menjamin kesinambungan sumber daya air Danau Singkarak. Unpublished thesis. Universitas Bung Hatta; 2018.
13. Ajiwibowo H, Ash-Shiddiq RHB, Pratama MB. Water quality and sedimentation modeling in singkarak lake, Western Sumatra. Int J GEOMATE. 2019;16:94–102. https://doi.org/10.21660/2019.54.8145.
14. Sumi F. Long-term change of water quality in Lake Singkarak observed from remote sensing data. IOP Conf Ser: Earth Environ Sci. Article in press. IOP Publishing. 2024;
15. Pusat Riset Limnologi dan Sumber Daya Air BRIN. Laporan teknis Danau Singkarak tahun 2021. Cibinong;
16. Novianti R, Yuniarti I, Ajie GS, Setiawan F, Triharyuni S, Afandi AY. System analysis and research priorities towards sustainable management of Lake Singkarak, Indonesia. IOP Conf Ser: Earth Environ Sci. IOP Publishing. 2023; https://doi.org/10.1088/1755-1315/1266/1/012022.
17. Amanda FF, Ghofur A. Studi Rekruitmen Dan Eksploitasi *Ikan bilih* Di Danau Singkarak Sumatera Barat. Proc Biol Edu Conf. 2016:701–3.
18. Kumala Sari P, Sujianto S. Implementasi Peraturan Gubernur Sumatera Barat Tentang Penggunaan Alat dan Bahan Penangkapan Ikan di Perairan Danau Singkarak. J Soc Policy Issues. 2021;3:104–9. https://doi.org/10.58835/jspi.v1i3.27.
19. Batubara J. Inventarisasi Alat Penangkap Ikan Di Danau Singkarak Provinsi Sumatera Barat. Article of Undergraduate. Unpublished. 2021. Universitas Bung Hatta. Available: https://ejurnal.bunghatta.ac.id/index.php/FPIK/article/view/5185.

Chapter 4
A Glimpse of the Decision Support System

Ivana Yuniarti and Satyo Gusti Anugrah

Abstract This chapter aims to provide a better understanding of the decision support system (DSS) before dwelling on the DSS silo models to give the readers clear pictures and a mind map of the following chapters. It elucidates the DSS structures and features, some output displays, processes to maintain the legitimacy of the framework and the data, and the efforts to ensure data equity and uncertainties. It also reflects on the scope and limitations of the DSS, the implications of the power of data, and possibly obscuring (further) uncertainties in the data itself.

Keywords Decision support systems · Silo model · Lake · Lake management · Data · Uncertainty

4.1 DSS Structure and the Composing Components

Following the EBM-DPSER framework (Chap. 3), the structure of the DSS addresses the management issues prioritized by stakeholders, which are:

1. Watershed land-use best practice management scenarios (composing models: Soil and Water Assessment Tool/SWAT, the aquatic-system model Water Ecosystems Tool (WET/QWET), and dose–response function for the economic valuation part).
2. *Ikan bilih* fishing regulation (constructed models: Stock growth algorithm, surplus production method, and market price method for the economic valuation part).

I. Yuniarti (✉)
Research Centre for Environmental Archaeology, Maritime Archaeology, and Cultural Sustainability, National Research and Innovation Agency (BRIN), Cibinong, Indonesia
e-mail: ivana.yuniarti@brin.go.id

S. G. Anugrah
Department of Informatics, Faculty of Industrial Technology, Universitas Atma Jaya Yogyakarta, Padang, Indonesia

K. Glenk et al. (eds.), *Decision Support Systems for Multisectoral Tropical Lake Management in Indonesia*, https://doi.org/10.1007/978-3-032-11268-2_4

3. Water inundation level management (used models: maximum entropy model, Geographical Information Systems (GIS) analysis, and damage compensation cost for the economic valuation).
4. Establishment of no-catch zone (applied models: maximum entropy model and replacement cost for the economic valuation).
5. Macroplastic management (composed model: contingent valuation method for the economic valuation).

4.2 DSS Features

The DSS is available at https://english-dss-singkarak.vercel.app/#/, with restricted access provided to Tanah Datar and Solok government agencies. It contains three main menus: the DSS, the scientific explanation for constructing the DSS, and default settings (Fig. 4.1). The figure also shows the DSS operation history.

The ***Decision Support Tools*** menu is an interactive interface where users can input the required data and run the system to obtain the outputs (Fig. 4.2). The management scenarios that are included in this menu are land-use best practice management scenarios (MS1), fishing effort regulation (MS2), water inundation level management (MS3), no-catch zone (MS4), and macroplastic control effort (MS5) (Fig. 4.2a–e). These management scenarios are selected to overcome stakeholders' prioritization of the lake's management issues by engaging in participatory methodology, as elucidated in Chap. 3. This procedure is one of our strategies to maintain the legitimacy of the DSS by incorporating stakeholders' perspectives while also dealing with experts' opinions. We use repetitive stakeholder consultations via several focus group discussions, interviews, expert meetings, and a stakeholder workshop (details in Chap. 3).

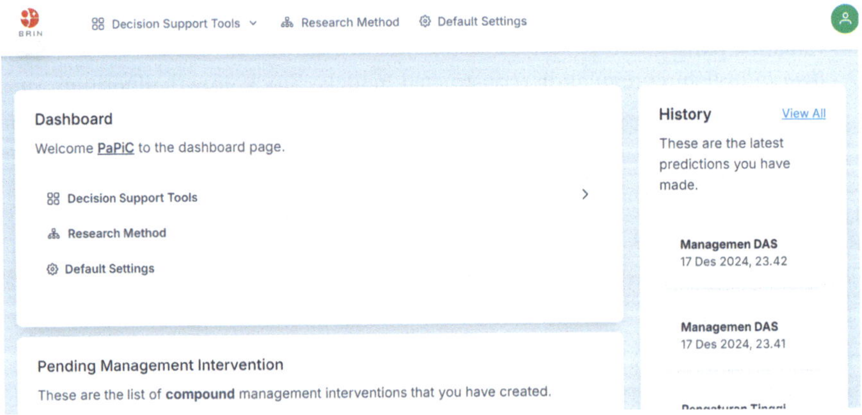

Fig. 4.1 The DSS main menu

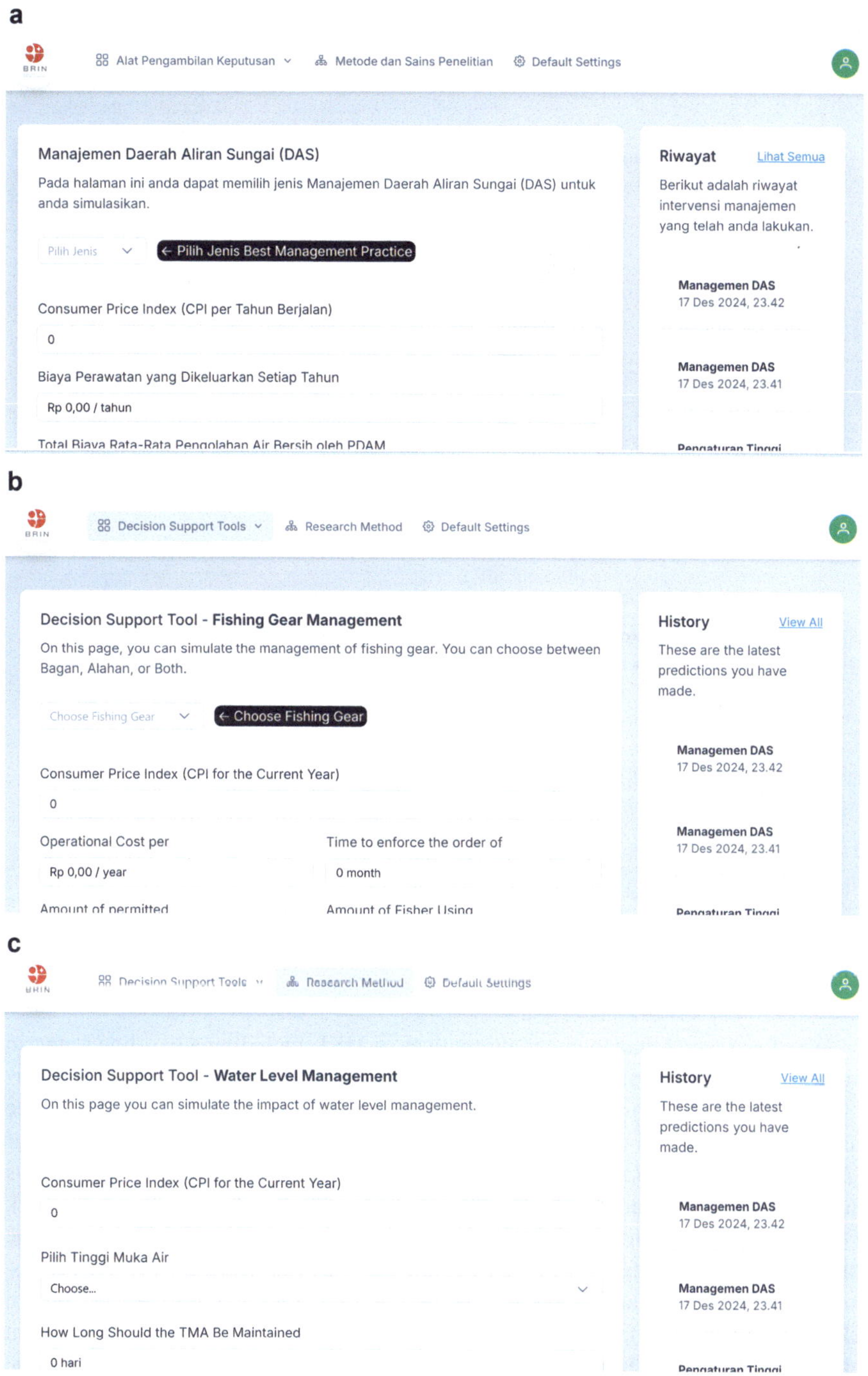

Fig. 4.2 User data input interface: (**a**) land-use management, (**b**) fishing effort regulation (MS2), (**c**) water inundation level management (MS3), (**d**) no-catch zone (MS4), (**e**) macroplastic control effort (MS5)

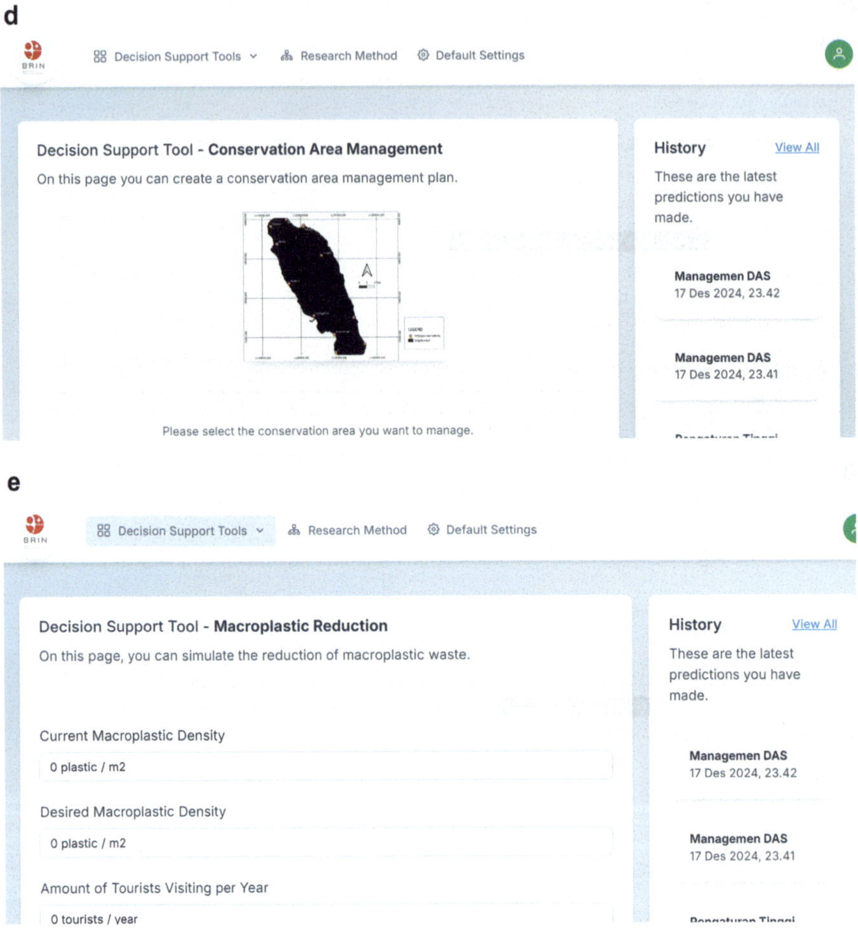

Fig. 4.2 (continued)

We offer several management alternatives in MS1, including a baseline scenario and tree planting, physical engineering best practice management (contouring, terraced field, and green belt), and a combination of physical engineering and tree planting (Fig. 4.3a). For MS2, we focus on regulating unsustainable fishing gear (*Bagan*-lift net and *Alahan*-a traditional trapping area) by creating new waterways and enclosing them, mostly combined with electrocuting and poisoning (Fig. 4.3b). For MS3, we provide five options for water level management: 361.5 m (the average factual water level obtained from five-year daily water level data); 362 m (the proposed water level by farmers); 362.5 (the minimum water level to support optimum electricity production); 363 m (the middle water level range to support optimum electricity production); and 363.5 m (the maximum water level assigned by the agreement between the local customary group and the hydropower

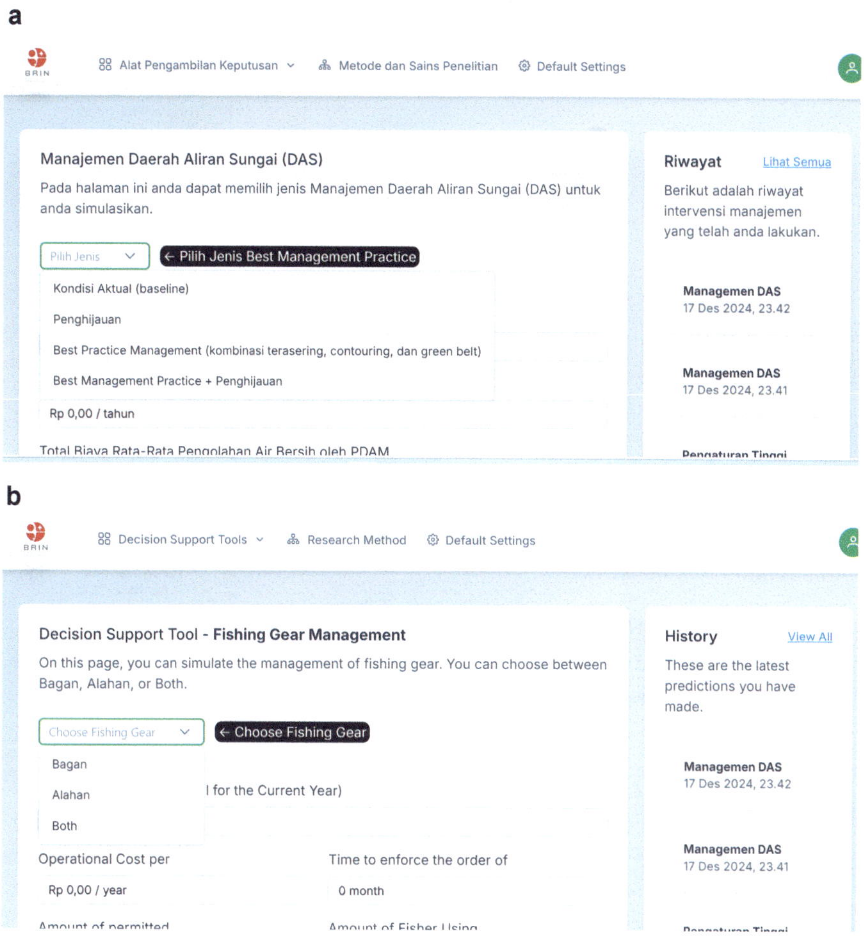

Fig. 4.3 Several management alternatives in each management response menu: (**a**) land-use management (MS1), (**b**) fishing effort regulation (MS2), (**c**) water inundation level management (MS3), (**d**) no-catch zone (MS4), (**e**) macroplastic control effort (MS5)

company—*Perusahaan Listrik Negara* or PLN Bukit Tinggi) (Fig. 4.3c). MS4 has two management alternatives: no-catch zone in the hotspot area and a non-hotspot area to incorporate different economic values of the management alternatives (Fig. 4.3d). Meanwhile, MS5 focuses on macroplastic management alternatives: adding more trash bins, adding more trash disposal schedules to the local dumpsters, trash removal from the lake, and recycling (Fig. 4.3e). These alternatives are selected based on local stakeholders' prioritization.

The outputs of the operation of the DSS are presented in three indicators: net present value (in millions 2024 IDR) and benefit-cost ratio (B/C)) to incorporate the decision-makers' interests in economic efficiency and the improvement of the sustainability index (SI) (Fig. 4.4). SI is selected since it is a composite index to

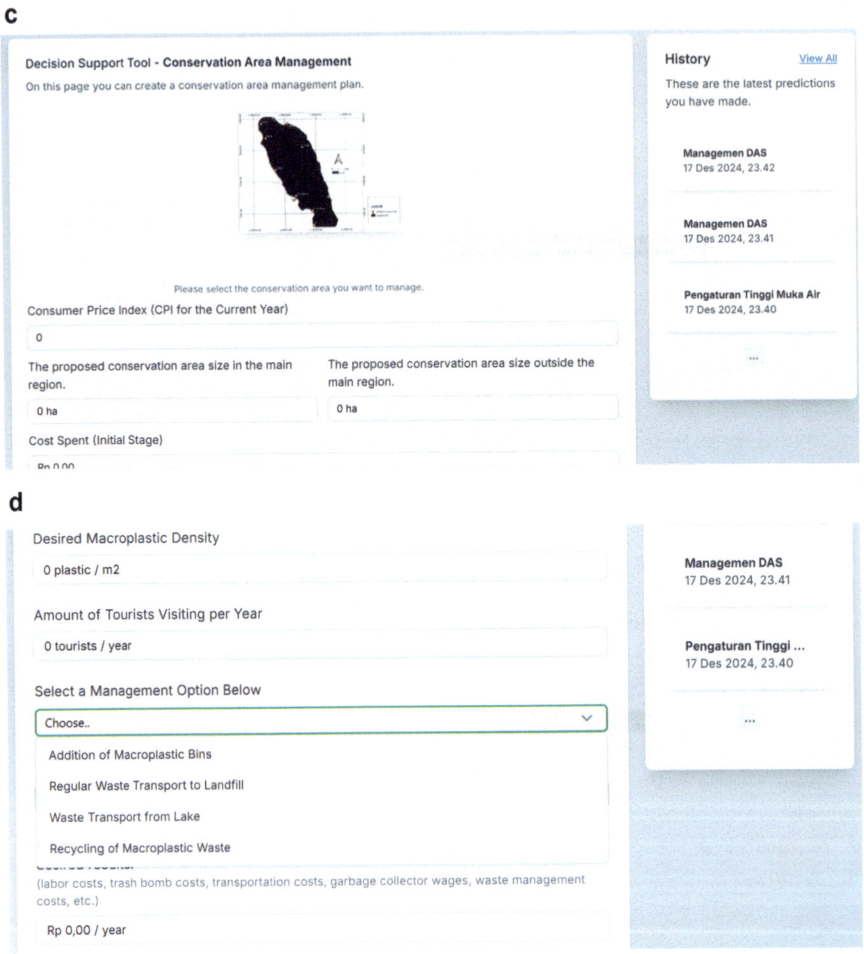

Fig. 4.3 (continued)

History Detail

Management Intervention	TSS	TN	10th-year Stock Projection (tons)	Habitat Area (ha)	Water Level	NPV in Million Rupiah	B/C	Change in SI
Pengurangan Makroplastik	-	-	-	-	-	-163 -146 -119	0.22 0.22 0.22	0.018

NPV: The net economic profit generated up to the final year of the simulation unit: rupiah. (If the NPV value is greater than 1, it means the intervention results in a positive profit).
BC ratio: The ratio of benefits to costs during the simulation years. (If the BC ratio is greater than 1, it means the intervention results in a positive profit).
SI: The sustainability index that reflects the impact of the management intervention on the sustainability of the lake.

Fig. 4.4 The basic outputs: Net Present Value (NPV), Benefit-Cost Ratio (B/C), and the improvement of sustainability index/SI

a

History Detail

Management Intervention	TSS	TN	10th-year Stock Projection (tons)	Habitat Area (ha)	Water Level	NPV in Million Rupiah	B/C	Change in SI
Managemen DAS	-	-	-	-	-	-34	0.00	0.028

NPV: The net economic profit generated up to the final year of the simulation unit: rupiah. (If the NPV value is greater than 1, it means the intervention results in a positive profit).
BC ratio: The ratio of benefits to costs during the simulation years. (If the BC ratio is greater than 1, it means the intervention results in a positive profit).
SI: The sustainability index that reflects the impact of the management intervention on the sustainability of the lake.

b

History Detail

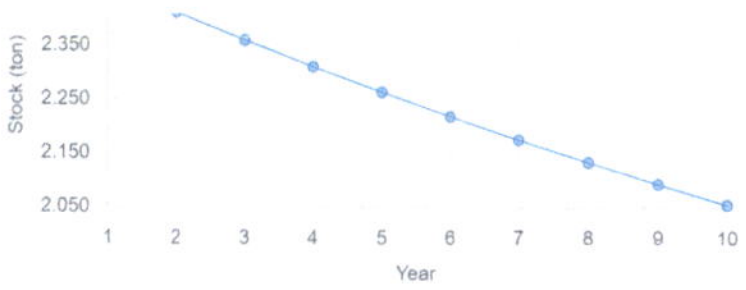

Management Intervention	TSS	TN	10th-year Stock Projection (tons)	Habitat Area (ha)	Water Level	NPV in Million Rupiah	B/C	Change in SI
Pengaturan Alat Tangkap Bagan	-	-	2.056,632	-	-	-2.135.769 -1.913.124 -1.511.035	0.10 0.11 0.14	0.055

c

History Detail

Management Intervention	TSS	TN	10th-year Stock Projection (tons)	Habitat Area (ha)	Water Level	NPV in Million Rupiah	B/C	Change in SI
Pengaturan Tinggi Muka Air	-	-	-	Luas Habitat (Hotspot): 623.24 Luas Habitat (Non-Hotspot): 1279.9 Luas Sawah Marjinal yang Terbanjiri: 33.51	362.5	42.297 37.725 30.717	56.42 56.42 56.42	0.033

NPV: The net economic profit generated up to the final year of the simulation unit: rupiah. (If the NPV value is greater than 1, it means the intervention results in a positive profit).
BC ratio: The ratio of benefits to costs during the simulation years. (If the BC ratio is greater than 1, it means the intervention results in a positive profit).
SI: The sustainability index that reflects the impact of the management intervention on the sustainability of the lake.

Fig. 4.5 Examples of output: (**a**) MS1, (**b**) MS2, (**c**) MS3

indicate the selected management scenario's contribution to improving lake ecosystem health. This is a self-tailored index constructed with the local stakeholders and expert panels' data and perspectives (more details in Chap. 8). The additional outputs for MS1 include total suspended solid (TSS) and total nitrogen (TN) (Fig. 4.5a) to inform the users of the forecasted water quality parameters impacted by the selected management scenario. For MS2, the additional output is predicted fish stock at the end of the operational year (10 years by default) and a graph of the

Management Intervention	TSS	TN	10th-year Stock Projection (tons)	Habitat Area (ha)	Water Level	NPV in Million Rupiah	B/C	Change in SI
Pengaturan Alat Tangkap Bagan	-	-	3.473,057	-	-	-62.276 -18.941 58.164	0.87 0.96 1.16	0.055
Pengurangan Makroplastik	-	-	-	-	-	3.211 2.864 2.332	4.84 4.84 4.84	0.018

Fig. 4.6 An example of compound output for MS2 and MS5

predicted stock in each operational year (Fig. 4.5b). For MS3, the additional output is the forecasted *ikan bilih*'s habitat extent area (hotspot and non-hotspot) and the extent area of the flooded rice field (Fig. 4.5c). Meanwhile, the obtained output (s) of each DSS operation can be displayed as a single management intervention or as a compound unit to simulate management scenario sets intended by the user (s) (Fig. 4.6).

The menu of ***Research Methods*****,** one of the core efforts to maintain data transparency, is a specific feature that describes the science and data behind the constructed DSS. It features an overview of Lake Singkarak's geological, morphological, and current water quality status (Fig. 4.7a), the basic information of the DPSIR and EBM-DPSER framework (Fig. 4.7b), the engaged EBM-DPSER framework to build the DSS (Fig. 4.7c), the description of the engaged models, the summary of the applied economic valuation methods, the used assumptions to build the models (Fig. 4.7d), and the description of cost-benefit analysis (CBA) and the SI (Fig. 4.7e).

The **default setting** menu is a tool for changing essential information, such as the time for management intervention, discount rate, and fish and labor price, which influence the whole system's operation. Users can customize the data, but it's best to consult each other when comparing simulation outputs.

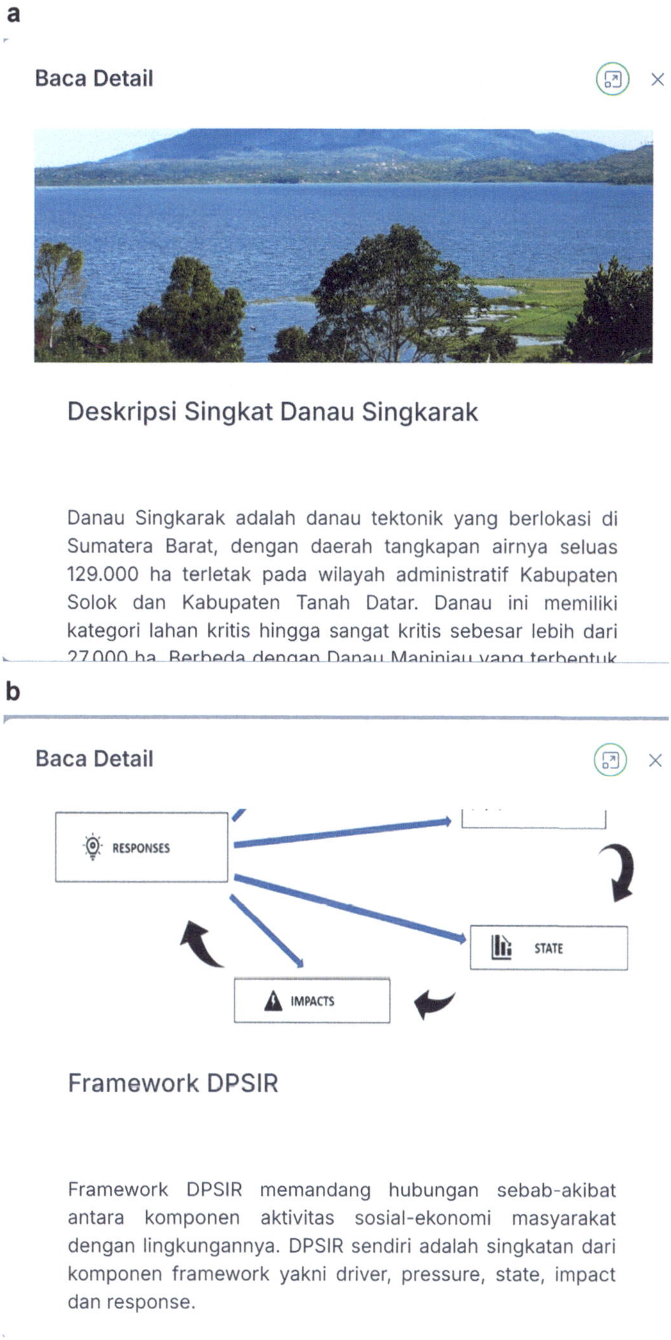

Fig. 4.7 Short information of (**a**) Lake Singkarak, (**b**) DPSIR framework, (**c**) the engaged EBM-DPSER framework, (**d**) applied models and economic valuation, (**e**) CBA and SI, (**f**) important data and their collection method

c

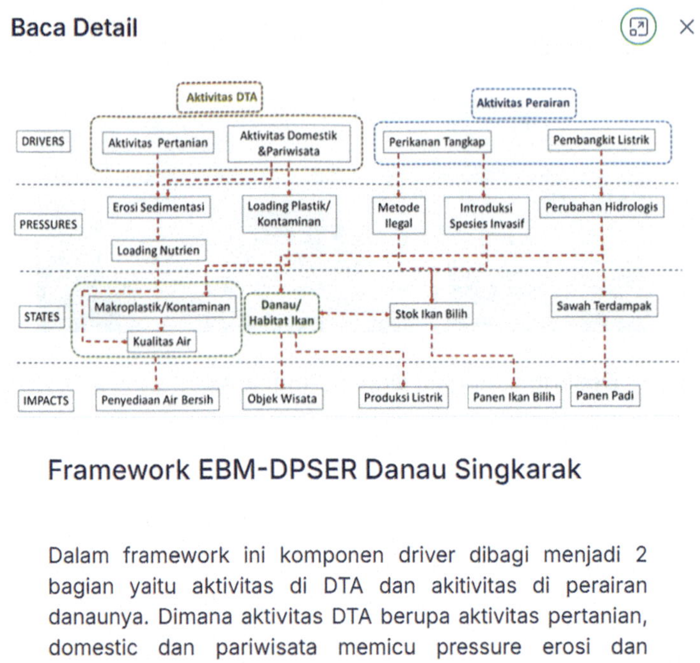

d

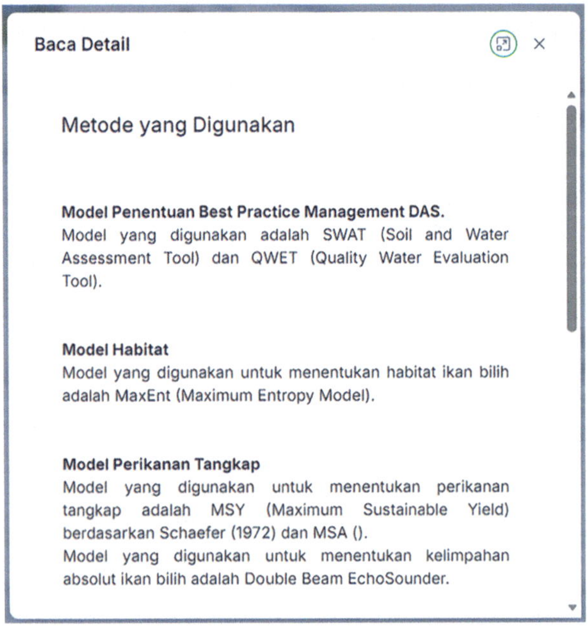

Fig. 4.7 (continued)

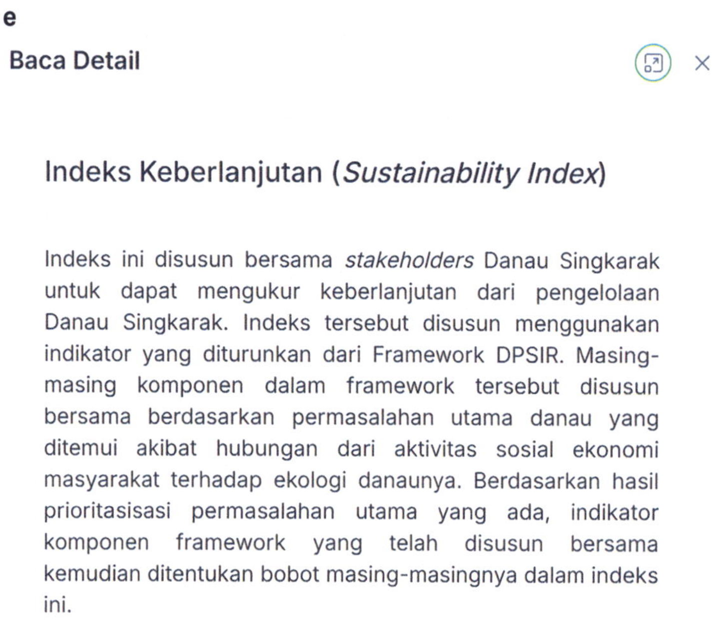

Fig. 4.7 (continued)

4.3 Scope and Limitation of the DSS

The DSS application is limited to Lake Singkarak due to its customization to overcome the specific management issues of the lake. Application of the DSS to other lakes is not advised considering the data used to build the silo models are specific (e.g., erosion rate, habitats) and the difference between each lake's system behavior.

We incorporate several uncertainties in each silo model (details in the following chapters), such as uncertain economic conditions, weather scenarios, and future temperature scenarios. However, we cannot include some sources of uncertainty in the analysis and the model constructions, such as prolonged extreme weather events, specific political situations (e.g., elections), and the unpredicted disruption of the biogeochemical cycles. We are aware that such situations occur, and they can affect data and decision-making, as we have incorporated in our first EBM-DPSER framework version (red circles in Fig. 4.8). However, limited time, labor, and budget obstructed our efforts to encompass such challenges.

In Chap. 2, we provide context and illustrate how such factors affect decision-making and the lake's condition. Thus, we expect that the readers will come to an equal understanding of what the DSS covers and what it does not, and they will understand our best efforts to incorporate them in our analysis. Other management

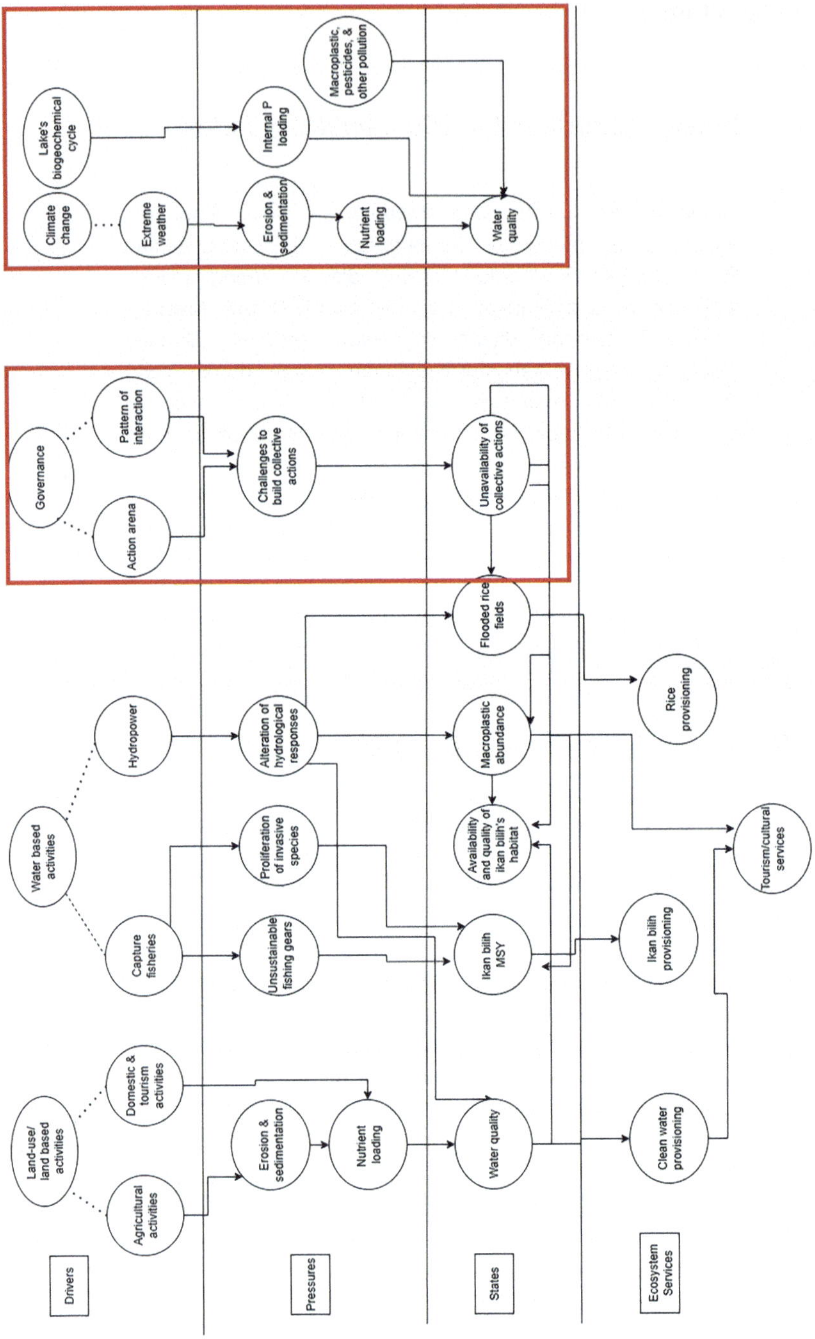

Fig. 4.8 Our first version of the EBM-DPSER framework. Red circles show the excluded part from the DSS

issues, such as stagnant tourist visits and invasive species, are not included in the DSS as they are not listed as the main stakeholders' concerns.

The DSS is specifically intended to answer decision-makers' needs; thus, it is only open to a limited set of users, consequently limiting other lake users, such as marginalized fishers, from accessing it. Initially, we aimed to make it open, but based on the request from the local government, it is currently distributed to specific stakeholders.

Such restrictions may deepen inequality in decision-making processes. We distribute access to the DSS to several local governmental agencies to reduce the inequalities by relying on the assumption that they can represent the marginalized stakeholders, which is a very optimistic assumption indeed. The following section discusses further related equity.

4.4 Data Legitimacy, Equity, Uncertainties, and Power of Data

As previously explained, we worked on several efforts to ensure that data legitimacy is considered in the DSS construction. In this section, firstly, we elaborate on our structural methods to overcome the challenges of data legitimacy by referring to the work of [1, 2], which includes transparency, bias mitigation, accountability frameworks, education and training, public engagement, and adaptive learning.

Transparency is thoroughly maintained by opening public data and listing the important data and assumptions, building a specific feature in the DSS for elaborating the methods, and conducting repeated stakeholder and expert consultative meetings in 2022–2023 to collect their perspectives, data, and priorities. We also publish each of our silo models and the overall DSS in international journals, proceedings, and books. We acknowledge that not all stakeholders agree on the framework and the data, as we encountered during the focus group discussion in the provincial planning agency. In such a case, we offered a specific discussion on long-term data series with the groups. Should the invited party agree to have the discussion, we continue with the plan. Otherwise, we continued with another expert meeting to discuss the data. In total, we arranged six series of interviews with local governments (*Nagari*–traditional village– and regency level), two series with fishers, two series of interviews with the hydroelectric company, two focus group discussions, and one workshop.

To maintain long-term data transparency, we share the DSS with various local institutions to update the data and settings, as they are the principal users of the DSS. Algorithmic bias is defined as systematically unfair results produced by algorithms [3] derived from several sources, such as data quality, modeling choices, and homogeneity of the development team. We mitigated the bias by exploring systematic bias mitigation [4] through the following means:

1. Maintaining an interdisciplinary team and expert panel of experts in fisheries, hydrology, GIS, remote sensing, lake management, stakeholder engagement, economic valuation, water quality, ecology, IT, modelers, and hydroacoustic.
2. Ensuring ethical standards are followed through data collection and processing. The ethical standard complied with and adhered to the National Research and Innovation Agency (BRIN) guideline obtained in Klirens Etik (brin.go.id).
3. Data adjustment: Data selection using specific statistical indicators (e.g., local spatial autocorrelation for the MaxEnt model) and long-term data series collection to represent seasonal data.
4. In-processing measurement: Data training includes separating training and test data, model tuning, field-data model validation, and publishing the model construction methods in international peer-reviewed scientific journals.
5. After processing measurements: The DSS relies on human-controlled and supervised data input to incorporate current updates and uncertainties into the system operations.
6. Stakeholder engagement: Stakeholder engagement since the beginning of the project (i.e., framework development), during the project (i.e., SI development), and after the project (training and discussions to obtain user feedback).

The powers of data refer to the significance of data in influencing stakeholders' decisions, strategies, and insights. The concept is related to data equity, which ensures fair and equal access to data and its benefits across different groups. It is relatively well addressed in developing the DSS and its subsequent activities. The data was collected from several sources (e.g., local and national governmental agencies, farmers, and fishers) and then opened and discussed in several focus group discussions. The selected data is stored in several online platforms (GitHub, BRIN Repository (RIN Data Repository (brin.go.id)) to ensure that stakeholders can access the data. However, as previously discussed, some stakeholder groups, such as farmers and fishers, can be further marginalized due to their limited access to online data sources. In such cases, we rely on their representativeness by local governmental agencies. In the long run, stakeholders can compare the outputs with data obtained from **SIDANAU**, an online Indonesian lake database, which can be accessed in SIDANAU—Sistem Informasi Danau.

Data uncertainties are handled in each silo model. The included sources of uncertainty are:

1. The varied data on initial fish stock, caused by the differences in the methodology used in secondary data sources, had substantial effects on the predicted stocks. To overcome this, we conducted a primary data survey using hydroacoustic (details in Chap. 6). After comparing the results, we provided two output scenarios based on primary and secondary data and relied on the stakeholders and expert panel's decisions on the more plausible outputs.
2. Weather conditions:
   (a) Its impacts on the fluctuations of turbidity and other water quality parameters are incorporated into the habitat model by including the generated long-

term water quality prediction scenarios in the development of the model. After that, we compared the outputs and identified general patterns. Ultimately, we selected the average values to build the final models, as they can represent the value ranges (details in Chap. 6).

(b) Uncertain weather conditions' effects on water level fluctuations are included in the habitat and flooded rice field models. We simulated six water inundation levels for habitat models to represent every 0.5 m increment of water level from 361 to 363.5 m. We selected this range to reduce the necessary spatial data processing. Higher resolution simulations using 20 cm increments did not yield significant differences; thus, we simulated inundation of rice fields for every 50 cm increment of water level.

3. Economic conditions. The uncertainty due to changes in economic situations is reflected in price changes and the consumer price index in this DSS. We accommodate the users' input for these two factors to provide the most updated information.
4. Tourist preference. Tourist preference was only included in the tourism and economic valuation model of macroplastic management. Considering data and time limitations, their preferences on water quality and land-use management were not included in the analysis.

Other uncertainties that can be important but are not included in the model construction are political situations, sudden change of governance type, alteration of the biogeochemical cycle, and climate change (Fig. 4.8).

References

1. Grimmelikhuijsen S, Meijer A. Legitimacy of algorithmic decision-making: six threats and the need for a calibrated institutional response. Perspect Public Manag Gov. 2022;5:232–42. https://doi.org/10.1093/ppmgov/gvac008.
2. Waldman A, Martin K. Governing algorithmic decisions: the role of decision importance and governance on perceived legitimacy of algorithmic decisions. Big Data Soc. 2022;9 https://doi.org/10.1177/20539517221100449.
3. Bagstad KJ, Villa F, Batker D, Harrison-Cox J, Voigt B, Johnson GW. From theoretical to actual ecosystem services: mapping beneficiaries and spatial flows in ecosystem service assessments. Ecol Soc. 2014;19 https://doi.org/10.5751/ES-06523-190264.
4. Barocas S, Hardt M, Narayanan A. Fairness and machine learning limitations and opportunities; 2023. p. 1–294. Available: https://fairmlbook.org/pdf/fairmlbook.pdf

Chapter 5
Decision Support Tool (1): Models of Impacts of Land Use Management on Water Quality

Iwan Ridwansyah, Aldiano Rahmadya, Atiqotun Fitriyah, Andi Muhammad Yasser Hakim, Kharistya Amaru, Edwin, Adi Witono, Anna Fadliah Rusydi, Sri Malahayati, Bambang Dwi Dasanto, Hari Jaya Sipayung, Hendro Wibowo, and Anastasia Lidya Eka Putri

Abstract This chapter presents the decision support tool (DST) for assessing the effects of land use changes on the hydrological response within Lake Singkarak's catchment area. We engaged in the integration of cellular automata (CA) and Markov chain (MC) models for forecasting land use changes over time to inform best management practices (BMPs) and spatial prioritization. The simulations of the models are one of the pivotal inputs for the final decision support systems (DSS). We briefly introduced the soil and water assessment tool (SWAT) to predict land use composition in Lake Singkarak's watershed from 1990 to 2020. The final outputs of change in water yield, the best management practice scenarios, and the validation methods and results are presented.

Keywords SWAT · Cellular Automata (CA) · Lake · Land use · Lake Singkarak · Best management practice · Soil and water · Hydrology

---

I. Ridwansyah (✉) · A. Rahmadya · A. Fitriyah · A. M. Y. Hakim · A. F. Rusydi · H. Wibowo
Research Centre for Limnology and Water Resources, National Research and Innovation Agency (BRIN), Cibinong, Indonesia
e-mail: iwan017@brin.go.id

K. Amaru · A. L. E. Putri
Agriculture Engineering Study Program, Padjadjaran University, Sumedang, Indonesia

Edwin
Agro-ecotechnology Department, Faculty of Agriculture, University of Andalas, Padang, Indonesia

A. Witono
Research Centre for Climate and Atmosphere, National Research and Innovation Agency (BRIN), Bandung, Indonesia

S. Malahayati · B. D. Dasanto · H. J. Sipayung
IPB University, Bogor, Indonesia

K. Glenk et al. (eds.), *Decision Support Systems for Multisectoral Tropical Lake Management in Indonesia*, https://doi.org/10.1007/978-3-032-11268-2_5

5.1 Chapter Introduction

This chapter presents the decision support tool (DST) for assessing the effects of land use changes on the hydrological response within Lake Singkarak's catchment area. As we will discuss, this DST helps identify and connect key environmental issues with practical policy or management interventions. This framework recognizes land use—such as agriculture, forestry, and urban expansion—as primary drivers that generate environmental pressures like erosion and nutrient loading. Environmental factors considered in the model include soil composition, topographic features, geological formations, and meteorological conditions. Human-induced factors include land use alterations and other anthropogenic activities, such as domestic, industrial, and agricultural waste disposal.

The interrelated environmental and human components eventually affect the lake's health, making it a crucial element of the watershed's hydrological system [1]. They modify the condition of the watershed, specifically water quality, which subsequently influences ecosystem services such as fishing production, drinkable water availability, and recreational value. Figure 5.1 depicts the chain of

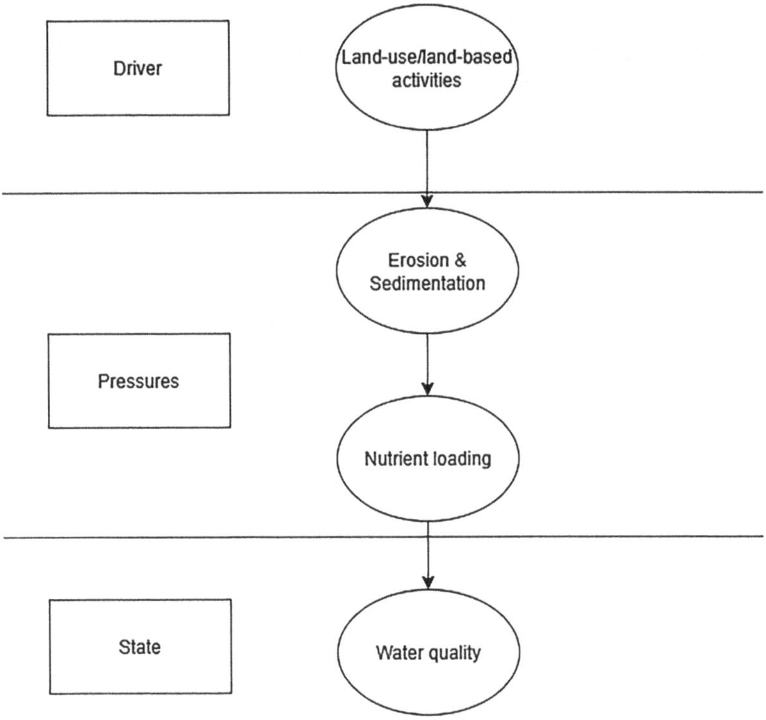

Fig. 5.1 Decision support tool in the EBM-DPSER framework

cause-and-effect linkages, highlighting the impact of terrestrial activities on watershed health via erosion, sedimentation, and nutrient loading.

5.2 Land Use/Land Cover Change Projection

Building on earlier chapters, which identified land use management as one of the key issues in Lake Singkarak management, this chapter focuses on how integrated modeling can inform best management practices (BMPs) and spatial prioritization. The core of the analysis involves simulating the impact of land use on watershed hydrology using historical and projected land use data. The integration of cellular automata (CA) and Markov chain (MC) models creates a sophisticated method for forecasting land use changes over time [2].

CA operates on a grid where each cell denotes a specific land use type, such as residential, commercial, agricultural, etc. [3]. Each cell's state changes based on interactions with neighboring cells, following defined transition rules that reflect the intricate dynamics of urban growth and how adjacent land uses affect each other [3]. Meanwhile, MC models augment this approach by offering a probabilistic framework that examines historical data to estimate the likelihood of transitions between various land use types [4]. This synergy enables a deeper understanding of land use evolution, in which CA illustrates the spatial distribution, and MC provides transition probabilities [5]. Together, they form a comprehensive modeling framework that enhances the accuracy of predictions of land use.

Predicting land use in TerrSet using CA and MC models consists of several key steps. We commenced by collecting and organizing historical land use data, which became the foundation for the analysis (in 2000 and 2020). Subsequently, a Markov Chain analysis is conducted to ascertain the transition probabilities across various land use categories throughout defined periods, yielding a quantitative comprehension of land use alterations. The CA model is subsequently constructed based on these probabilities, encompassing the grid structure, initial conditions, and local transition rules. The CA simulation is utilized to project future land use scenarios for 2020, with the results presented visually for assessment. Ultimately, to ensure precision and enhance the model for subsequent applications, the outcomes are evaluated against documented land use alterations (projected 2020 and actual 2020). Our research seeks to assess land use circumstances in 2030 and 2040.

5.3 Soil and Water Assessment Tool (SWAT) Model as a Tool for Land Use Management

The Soil and Water Assessment Tool (SWAT) represents a semi-distributed continuous model proficient in simulating long-term processes, rendering it a significant asset for watershed management [6]. This application can be implemented across various Geographic Information System (GIS) platforms, including ArcGIS and QGIS. Numerous investigations have utilized this model to evaluate water balance in response to land use alterations, functioning as a decision-making instrument for spatial planning [6–8]. The SWAT model, version 1.5.2, employed in this work, is a tool/plugin incorporated within the open-source GIS application—QGIS 3.16 Hannover.

The 2020 land use data is the benchmark for calibration and validation processes. Following a comprehensive validation of the SWAT model (Fig. 5.2) based on Moriasi et al.'s criteria for hydrological models [9], land use datasets from 1990, 2000, and 2010, as well as projected land use for 2030 and 2040, were incorporated to evaluate the model's response to various hydrological components, including surface runoff, interflow, base flow, and the maximum-minimum discharge ratio. The SWAT model for Lake Singkarak employs rainfall data sourced from the Climate Hazard Group Infrared Precipitation with Station (CHIRPS), calibrated using local precipitation measurements, enabling simulations from 1981 to July 2024. The model parameters are established using 30 × 30 m resolution data. At the same time, soil maps are produced from global Food and Agriculture Organization (FAO) soil data, with texture and structure assessed through laboratory soil sampling. The

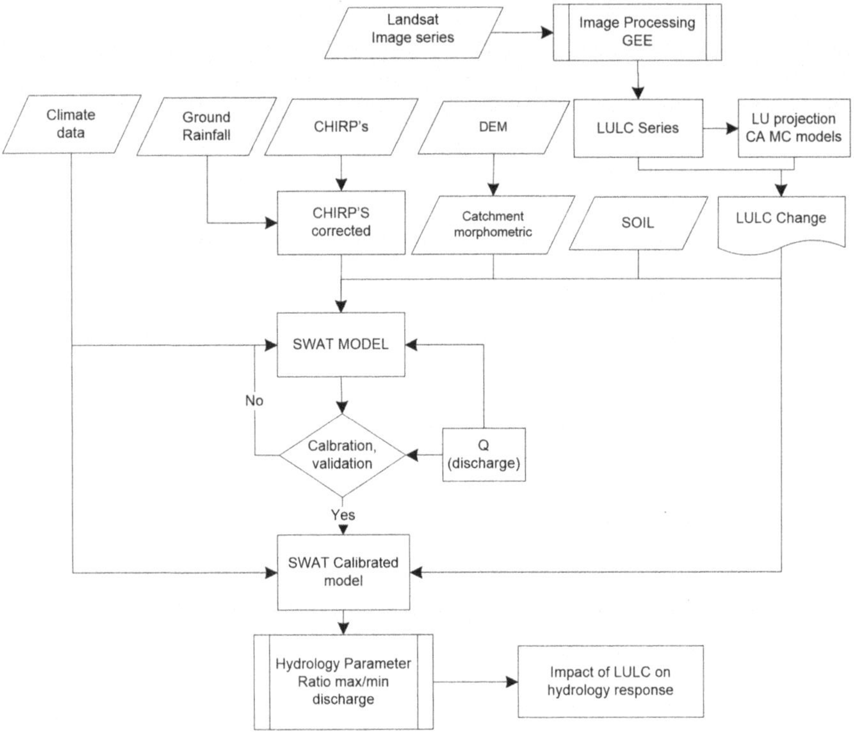

Fig. 5.2 Flowchart of model integration

SWAT model was calibrated and validated through manual and automated procedures. This method involved the analysis of outcomes through observational data, employing the Nash and Sutcliffe equation [10] and the coefficient of correlation (R^2) as categorized by [9, 11, 12]. The data for assessing the model's reliability included discharge observations from West Sumatra's Water Management Center (*Pengairan dan Sumber Daya Air Pekerjaan Umum* or PSDA-PU) at the Sumani and Sumpur Rivers.

5.4 Land Use Change Scenario

It is important to assess the model's performance and ability to accurately represent land use changes. The accuracy evaluation (Fig. 5.3) relies on several key metrics: Kstandar, Kno, Klocation, and KlocationStrata. Notably, all these metrics demonstrate values exceeding 70%, indicating a high level of agreement between the predicted and actual land use data. These impressive figures strongly suggest that the model is reliable and well-suited for future land use change predictions.

The land use in Lake Singkarak's watershed has undergone significant change from 1990 to 2020 (Table 5.1 and Fig. 5.4), particularly with the expansion of forest areas, which increased from 20.45% to 37.91%, reflecting efforts in reforestation and environmental conservation. In contrast, plantation areas saw a notable decline, dropping from 32% in 1990 to just 19% by 2020, indicating potential shifts in agricultural practices and economic priorities. Similarly, agricultural land area was reduced from 28.93% to 14.81% during the same period, highlighting a complex transformation in land use dynamics within the watershed and the interplay between environmental management and agricultural development.

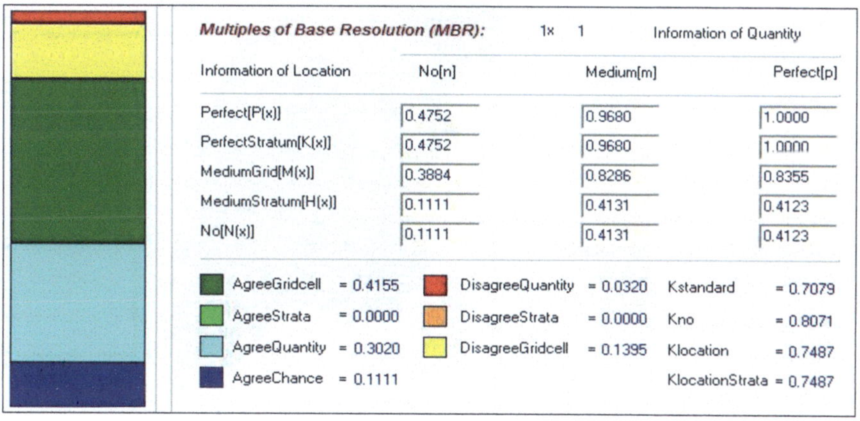

Fig. 5.3 Validation results comparing the actual land use of 2019 with the predicted land use for that year

Table 5.1 Land use composition in Lake Singkarak's watershed from 1990 to 2020

No	Land use	1990		2000		2010		2020	
		Ha	%	Ha	%	Ha	%	Ha	%
0	No data	1	0.00	0	0.00	186	0.16	234	0.20
1	Waterbody	12,033	10.54	12,025	10.53	11,920	10.44	12,057	10.56
2	Forest	23,345	20.45	27,453	24.05	35,274	30.90	43,286	37.91
3	Plantation	34,595	30.30	24,112	21.12	21,516	18.85	14,301	12.53
4	Open area	156	0.14	146	0.13	748	0.66	315	0.28
5	Built-up area	1294	1.13	1720	1.51	2266	1.98	4805	4.21
6	Paddy field	33,026	28.93	26,411	23.13	16,706	14.63	16,909	14.81
7	Shrub	604	0.53	15,527	13.60	12,272	10.75	9205	8.06
8	Moor	9116	7.98	6776	5.94	13,282	11.63	13,058	11.44
Total		114,170	100.00	114,170	100.00	114,170	100.00	114,170	100.00

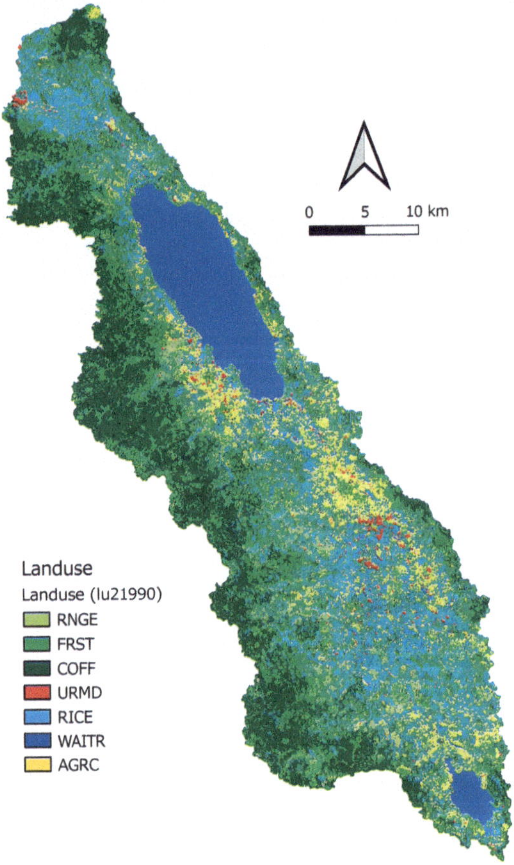

Fig. 5.4 Land use series in Singkarak Watershed from 1990 to 2020

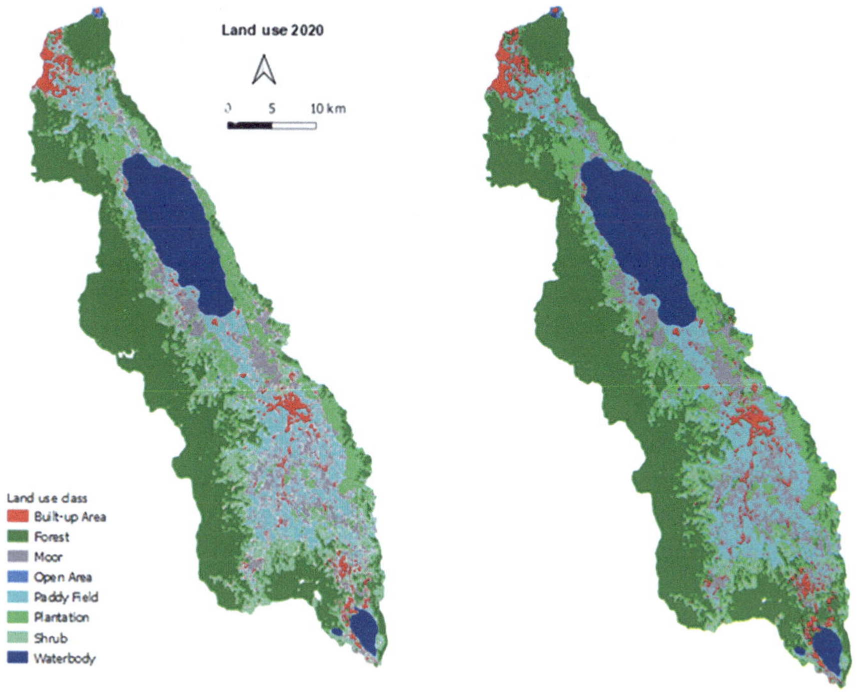

Fig. 5.5 Land use projection in 2030 and 2040

Table 5.2 Land use composition in Lake Singkarak's watershed from predicted years of 2030 to 2040

No	Land use	2030		2040	
		Ha	%	Ha	%
0	No data	234	0.20	234	0.20
1	Waterbody	12,089	10.59	12,123	10.62
2	Forest	48,692	42.65	52,404	45.90
3	Plantation	12,049	10.55	11,036	9.67
4	Open area	226	0.20	222	0.19
5	Built-up area	5717	5.01	5932	5.20
6	Paddy field	15,726	13.77	14,502	12.70
7	Shrub	8102	7.10	7436	6.51
8	Moor	11,335	9.93	10,281	9.00
Total		114,170	100.00	114,170	100.00

The future land use projection was simulated to effectively generate watershed conditions aligning with various hydrological parameters, providing insights into how land use changes might impact water resources. The land use maps for the years 2030 and 2040 highlight these anticipated changes (Fig. 5.5 and Table 5.2).

Forests are projected to remain the dominant land cover type in 2030, encompassing 48,692 ha, which constitutes 42.65% of the total watershed area. This trend is expected to continue, with forested areas anticipated to increase to 52,404 ha by 2040, representing 45.90% of the watershed. This growth in the forest area suggests a positive trend towards preserving and expanding natural ecosystems.

Conversely, the built-up area is also projected to experience growth, rising from 5717 ha, or 5.01% of the total area in 2030, to 5932 ha, which equates to 5.20% by 2040. This increase in built-up land indicates a gradual conversion of other land uses into developed spaces, reflecting ongoing urbanization and land use change within the watershed. These dynamics underscore the need for careful planning and management to balance ecological health with developmental pressures.

5.5 Model Setup and Calibration

The SWAT simulation model requires additional data inputs and set up specifications to be defined. River discharge and precipitation data were sourced from the Sumatra River Basin Authority V. Land use information was acquired by classifying Landsat images employing Google Earth Engine (GEE). In contrast, supplementary data was amassed through field measurements and observations.

The preliminary stage of the SWAT hydrological modeling procedure is called Automatic Watershed Development (AWD). This stage generates a semi-distributed model structure by utilizing digital elevation modeling (DEM) data, with a threshold area of 1000 ha, delineating 120 sub-watersheds. The subsequent phase entails the definition of Hydrology Response Units (HRUs), employing land use data from 2020 as the simulation baseline, in conjunction with FAO soil data and slope classifications categorized into five distinct categories. This process culminates in the identification of 58 HRUs. Using the inputs (Table 5.3), the simulation process was initiated. Following the initial simulation, the model underwent calibration and validation to ensure accuracy in representing the watershed's hydrological behavior represented by parameters (Table 5.4).

The SWAT hydrological model was calibrated and validated at two monitoring stations operated by Balai Besar Wilayah Sungai/BBWS Sumatra V–Major River Basin Organization Sumatra V, specifically at the Sumani and Malalo discharge stations. The results from these two locations showed notable differences, with the Sumpur River exhibiting better criteria than the Sumani River. The calibration results at the Malalo station on the Sumpur River yielded the coefficient of determination (R^2) and the Nash–Sutcliffe Efficiency (NSE) values of 0.56 and 0.55 for daily simulations. In contrast, the monthly simulations reached values of 0.89 and 0.77 (Fig. 5.6 and Table 5.5).

Table 5.3 Summary of data requirements

No	Data	Data format	Sources
1	Digital elevation model (DEM)	Grid (cell size 30 × 30 m)	Shuttle Radar Topography Mission (SRTM) of USGS
2	Soil map scale 1:250,000	Vector map(.shp)	Global data FAO
3	Meteorological data (2005–2008)	Table (text)	Meteorological, Climatological, and Geophysical Agency CHIRPS (corrected by ground rain gauge)
4	Landsat image (1990, 1997, 2005 and 2015)	Raster	Landsat, United States Geological Survey (USGS) and ground check
5	Soil properties for SWAT database	Numeric	Field sampling and laboratory analysis
6	Daily discharge	Table (.dbf and . txt)	Ministry of Public Works and Housing, field data

Table 5.4 Optimum calibration parameters for streamflow and sediment model

	No	Parameter name	Description	Fitted value	Min value	Max value
Stream flow parameters	1	r_CN2.mgt	SCS-CN for moisture condition II	−0.10	−0.2	0.2
	2	v_ALPHA_ BF.gw	Base flow alpha factor (1/days)	0.73	0	1
	3	v_GW_ DELAY.gw	Groundwater delay time (days)	42.6	30	450
	4	v__ GWQMN. gw	Threshold depth of water in the shallow aquifer required for return flow to occur (mm H_2O)	1.26	0	2
	5	v_GW_ REVAP.gw	Groundwater revap coefficient	0.028	0	0.2
	6	v_ESCO.hru	Soil evaporation compensation factor	0.89	0.8	1
	7	v_CH_ N2.rte	Manning's "*n*" value for the main channel	0.12	0	0.3
	8	v_CH_ K2.rte	Effective hydraulic conductivity in main channel alluvium (mm/h)	233.7	5	500
	9	v_ALPHA_ BNK.rte	Base flow alpha factor for bank storage (days)	0.89	0	1
	10	r_SOL_ AWC.sol	Soil available water storage	10.7	−15	15
	11	r_SOL_K. sol	Soil hydraulic conductivity	−5.85	−8	8
	12	r_SOL_ BD.sol	Soil bulk density	1.4	−0.5	2

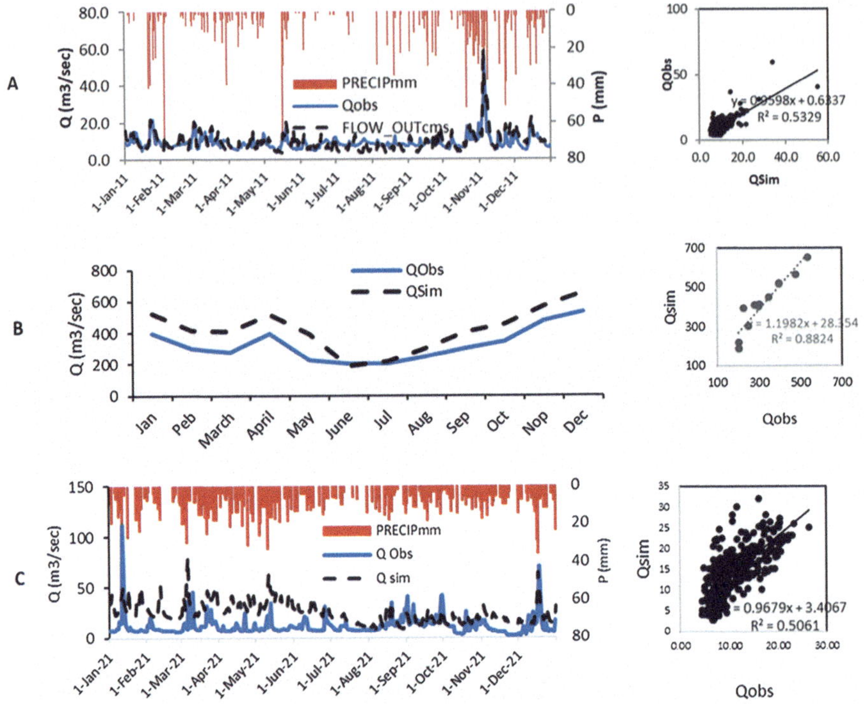

Fig. 5.6 Daily simulation of river flow (black line) and field observation data of precipitation rate (red line) and river flow (blue line) at (**a**) Malalo station, (**b**) Sumani station, and (**c**) correlation of monthly data

Table 5.5 Result of model validation in Lake Singkarak's watershed

Location	Resolution	Coefficient of determination (R^2)		Nash–Sutcliffe model efficiency (NSE)	
		Calibration	Validation	Calibration	Validation
Sumani R.	Daily	0.54	0.51	0.52	0.50
Sumani R.		0.56	0.53	0.55	0.51
Sumani R.	Monthly	0.79	0.73	0.64	0.61
Sumani R.		0.89	0.88	0.77	0.71

5.6 Impact of Land Use Change on Water Yield

The simulation results (Table 5.6) show a variation in the values of water yield parameters. Specifically, surface runoff increased from 483.9 mm/year in 1990 to 464.1 mm/year in 2000, followed by a decline to 431.8 mm/year in 2010, and a further decrease to 400.1 mm/year by 2020. In contrast, the base flow parameters exhibited a distinct trend, showing a decrease from 588.4 mm/year in 1990 to 537.6 mm/year in 2000, followed by an increase to 600.3 mm/year in 2010, and

Table 5.6 Hydrological model simulation results on land use in 1990, 2000, 2010, and 2020

Hydrology parameters		Land use 1990 mm/year	%	Land use 2000 mm/year	%	Land use 2010 mm/year	%	Land use 2020 mm/year	%
Surface runoff (mm/year)		483.9	18.7	464.1	18.3	431.8	17.2	400.1	15.9
Lateral flow (mm/year)		580.6	22.5	627.6	24.8	568	22.6	623.4	24.8
Base flow (mm/year)		674.8	26.1	600.5	23.7	668.7	26.6	676.6	26.9
Epa. And transpiration		809.4	31.3	807.2	31.1	809.8	30.4	809.9	29.9
Recharge to deep aquifer (mm/year)		36.7	1.4	32.8	1.3	36.4	1.4	37.8	1.5
Precipitation		2803.2		2803.2		2803.2		2803.2	
CN average		79.5		79.2		78.7		78.4	
Sumpur R.	Min Q (m^3/days)	1		0.4		1.1		2.8	
	Max Q (m^3/days)	27		17.3		24		34	
	Max Q/Min Q	34.2		39.6		22.7		12.4	
Sumani R.	Min Q (m^3/days)	3.8		2.9		5		6.9	
	Max Q (m^3/days)	129		125.6		127.4		82.8	
	Max Q/Min Q	34.2		43.9		25.3		12	

subsequently a decline to 592.1 mm/year in 2020. A similar pattern of change was observed in the lateral flow parameters, which recorded values of 515.5 mm/year, 575.6 mm/year, 521.1 mm/year, and 543.3 mm/year, respectively.

The impact of land use changes is also assessed through the analysis of the daily maximum-minimum discharge ratios of two major rivers that flow into Lake Singkarak: the Sumpur River (sub-watershed no. 26) and the Sumani River (sub-watershed no. 11). In line with the water balance composition, the maximum-minimum discharge ratios show a similar trend, increasing from the early 1990s to 2000, with the Sumpur River and the Sumani River rising from 34.2 to 39.6. Subsequently, from 2000 to 2020, both rivers' maximum-minimum discharge ratios declined. Additionally, the maximum/minimum discharge ratios were analyzed based on projected land use for the years 2030 and 2040 (Table 5.7), yielding values of 20.4 and 20.3 for the Sumpur River, while the Sumani River showed values of 31.4 and 25.3.

With the trend of increasing forest area in the Lake Singkarak watershed—from 20.45% in 1990 to 37.91% in 2020 and projected to reach 45.90% by 2040—there will be significant positive impacts for the community and the environment in the region. This expansion of forest land not only enhances the ecological balance [13] but also provides many benefits that can improve the quality of life for residents [14]. Forests contribute to soil enrichment by decomposing organic matter and the

Table 5.7 Hydrological model simulation results on land use projection in 2030 and 2040

Hydrology par.		Land use 2020 mm/year	%	Land use 2040 mm/year	%
Surface runoff (mm/year)		449.8	17.3	497.3	20.3
Lateral flow (mm/year)		545.5	21	575.6	23.5
Base flow (mm/year)		755.1	29	537.6	20
Epa. and transpiration		808.2	31.1	807.2	33
Recharge to the deep aquifer (mm/year)		41	1.6	29.5	1.2
Precipitation		2803.2		2803.2	
CN average		79.5		79.2	
Sumpur R	Min Q (m^3/days)	2.3		2.3	
	Max Q (m^3/days)	46		46.2	
	Max Q/Min Q	20.4		20.3	
Sumani R.	Min Q (m^3/days)	4.2		5.1	
	Max Q (m^3/days)	131.4		130	
	Max Q/Min Q	31.4		25.3	

nutrient cycling facilitated by the diverse plants that thrive within them [15]. The rich biodiversity in these ecosystems supports various species, providing essential resources and services that benefit human communities, such as food, medicine, and raw materials [16].

Moreover, the increasing forest cover adds to the region's natural wealth [16], offering opportunities for sustainable economic development through eco-tourism and responsible forestry practices [17]. Importantly, these forested areas play a vital role in disaster prevention [18]; they act as natural barriers against floods and landslides, helping to stabilize the soil and absorb excess rainfall [19]. In doing so, they protect local communities and contribute to climate change mitigation by sequestering carbon dioxide and regulating local climates [20]. Overall, the growth of forest areas in Lake Singkarak's watershed signifies a positive step towards ecological resilience and sustainable development, benefiting both the environment and the people who depend on it. Furthermore, expanding forested areas in the lake's watershed will enhance the availability of water-related ecosystem services.

The improvement of the lake's watershed is attributed to the expansion of forest cover, resulting in enhanced watershed conditions where a greater volume of rainfall is absorbed into the soil [21, 22]. However, projections for land use in 2030 and 2040 indicate that despite the increase in forest area, there will also be a rise in developed land and agricultural areas due to the growing population's demand for residential space. This trend is expected to lead to an increase in surface runoff and a decrease in groundwater infiltration. This observation aligns with research conducted on the Haihe River, which found that vegetation restoration reduces surface runoff while increasing base flow [23].

Conservation efforts by the West Sumatra Forestry and Environment Agency have produced significant benefits for water resources, evidenced by the decreasing ratio of maximum to minimum discharge, a key parameter for assessing watershed health in Indonesia. In 1990, the max-min discharge ratio was recorded at 26.8 for the Sumpur River and 34.2 for the Sumani River, which increased to 39.6 and 43.9, respectively, by 2000. Subsequently, this ratio declined in 2010 and 2020 and in the projected land use simulations for 2030 and 2040. Improving forest cover quality impacts local scales and contributes to the global water balance [24, 25]. In contrast, the situation in the Batanghari watershed, where deforestation has led to deteriorating conditions characterized by increased surface runoff, erosion, and reduced base flow, highlights a different outcome [26].

5.7 Contribution to the Decision Support System

The simulation results on land use change, particularly the increase in forest cover and its hydrological implications, serve as critical inputs for the Decision Support System (DSS) developed for the lake's watershed. By quantifying the effects of forest expansion and built-up area growth on surface runoff, base flow, and the max-min discharge ratio, these findings inform the system's capacity to evaluate different land management scenarios. The DSS uses the outputs from the SWAT model to compare alternative future land use trajectories and to assess their impact on water-related ecosystem services. This enables policymakers to visualize trade-offs between conservation, development, and hydrological sustainability. Moreover, the spatial identification of areas at risk of deforestation, especially in the southern sub-watershed, provides a targeted basis for local government interventions, such as zoning regulations or reforestation programs. As such, this integrated analysis strengthens the DSS's role in supporting evidence-based planning, enhancing resilience, and aligning land use decisions with sustainable watershed management goals.

References

1. Weyhenmeyer GA, Bogard MJ, Couture RM, de Wit HA, Frassl MA, Grabs T, Zarnetske JP. Global lake health in the Anthropocene: societal implications and treatment strategies. Earth's Future. 2023;11(6):e2023EF004387. https://doi.org/10.1029/2023EF004387.
2. Beyene E, Minale AS. Modeling urban land use dynamics using Markov-chain and cellular automata in Gondar City, Northwest Ethiopia. Chin J Popul Resour Environ. 2023;21:111–20. https://doi.org/10.1016/j.cjpre.2023.06.007.
3. Hadj-Ali C, Roy D, Amireche L, Antoni JP. Development of a cellular automata-based model approach for sustainable planning of affordable housing projects: an application case study in Algiers. Land Use Policy. 2023;125. https://doi.org/10.1016/j.landusepol.2022.106468.
4. Yang X, Zheng XQ, Chen R. A land use change model: integrating landscape pattern indexes and Markov-CA. Ecol Model. 2014;283:1–7. https://doi.org/10.1016/j.ecolmodel.2014.03.011.
5. Yasir M, Hui S, Binghu H, Rahman SU. Coastline extraction and land use change analysis using remote sensing and GIS—a review of the literature. Rev Environ Health. 2020;35:453–60. https://doi.org/10.1515/reveh-2019-0103.
6. Fohrer N, Haverkamp S, Eckhardt K, Frede HG. Hydrologic response to land use changes on the catchment scale. Phys Chem Earth Part B Hydrol Oceans Atmos. 2001;26(7–8):577–82. https://doi.org/10.1016/S1464-1909(01)00052-1.
7. Ridwansyah I, Yuliati M, Apip A, Onodera S, Shimizu Y, Wibowo H, et al. The impact of land use and climate change on surface runoff and groundwater in Cimanuk Watershed, Indonesia. Limnology. 2020;21(3):487–98. https://doi.org/10.1007/s10201-020-00629-9.

8. Ridwansyah I, Yuliati M, Wibowo H. Soil Water Analysis Tools (SWAT) hydrology modelling as a basis for spatial planning: a case study in Cimandiri Watershed, West Java Province. IOP Conf Ser Earth Environ Sci. 2019;380:012017. https://doi.org/10.1088/1755-1315/380/1/012017.
9. Moriasi DN, Arnold JG, Van Liew MW, Bingner RL, Harmel RD, Veith TL. Model evaluation guidelines for systematic quantification of accuracy in watershed simulations. Trans ASABE. 2007;50:885–900. https://doi.org/10.13031/2013.23153.
10. Nash JE, Sutcliffe JV. River flow forecasting through conceptual models part I—a discussion of principles. J Hydrol. 1970;10(3):282–90. https://doi.org/10.1016/0022-1694(70)90255-6.
11. Krause, P., Boyle, D. P., and Bäse, F.: Comparison of different efficiency criteria for hydrological model assessment, Adv. Geosci. 2005;(5):89–97. https://doi.org/10.5194/adgeo-5-89-2005.
12. Oeurng C, Sauvage S, Sánchez-Pérez J. Assessment of hydrology, sediment and particulate organic carbon yield in a large agricultural catchment using the SWAT model. J Hydrol. 2011;401(1–2):145–53. https://doi.org/10.1016/j.jhydrol.2011.02.017.
13. Gunawan H, Mulyanto B, Suharti S, Subarudi S, Ekawati S, Karlina E, et al. Forest land redistribution and its relevance to biodiversity conservation and climate change issues in Indonesia. For Sci Technol. 2024;20:213–28. https://doi.org/10.1080/21580103.2024.2347902.
14. Hallaj Z, Bijani M, Karamidehkordi E, Yousefpour R, Yousefzadeh H. Forest land use change effects on biodiversity ecosystem services and human well-being: a systematic analysis. Environ Sustain Indic. 2024;23:100445. https://doi.org/10.1016/j.indic.2024.100445.
15. Mangueira JRSA, Holl KD, Rodrigues RR. Enrichment planting to restore degraded tropical forest fragments in Brazil. Ecosyst People. 2019;15:3–10. https://doi.org/10.1080/21513732.2018.1529707.
16. Molnar JL, Kubiszewski I. Managing natural wealth: research and implementation of ecosystem services in the United States and Canada. Ecosyst Serv. 2012;2:45–55. https://doi.org/10.1016/j.ecoser.2012.09.005.
17. Zainal S, Nirzalin N, Fakhrurrazi F, Yunanda R, Ilham I, Badaruddin B. Actualizing local knowledge for sustainable ecotourism development in a protected forest area: insights from the Gayonese in Aceh Tengah, Indonesia. Cogent Soc Sci. 2024;10. https://doi.org/10.1080/23311886.2024.2302212.
18. FAO and UNEP. The state of the world's forests 2020: forests, biodiversity, and people. Rome. 2020, https://doi.org/10.4060/ca8642en.
19. Bhattacharjee K, Behera B. Forest cover change and flood hazards in India. Land Use Policy. 2017;67:436–48. https://doi.org/10.1016/j.landusepol.2017.06.013.
20. Njana MA, Chamshama SAO, Zahabu E, Malimbwi RE. Forest ecosystem services and their role in climate change mitigation and adaptation in Tanzania. Clim Dev. 2021;13(2):133–45. https://doi.org/10.1080/17565529.2020.1771253.
21. Pawitan, H. 2004. Perubahan penggunaan lahan dan pengaruhnya terhadap hidrologi Daerah Aliran Sungai (LandUse Changes and Their Impacts on Watershed Hydrology). IPB-Bogor. Bogor. Indonesia. Available https://www.academia.edu/28164908/PERUBAHAN_PENGGUNAAN_LAHAN_DAN_PENGARUHNYA_TERHADAP_HIDROLOGI_DAERAH_ALIRAN_SUNGAI_Land_Use_Changes_and_Their_Impacts_on_Watershed_Hydrology.
22. Chang Y, Liu X, Cao B, Wang Y, Shen YJ. Effects of vegetation restoration on runoff and its components in the mountainous Haihe River Basin. J Hydrol Reg Stud. 2024:101803. https://doi.org/10.1016/j.ejrh.2024.101803.
23. Jones JA, Wei X, Archer E, Bishop K, Blanco JA, Ellison D, et al. Forest-water interactions under global change. In: Forest-water interactions. Ecological studies 240; 2020. https://doi.org/10.1007/978-3-030-26086-6_24.
24. Wangpimool W, Pongput K, Sukvibool C, Sombatpanit S, Gassman PW. The effect of reforestation on stream flow in Upper Nan river basin using Soil and Water Assessment Tool (SWAT) model. Int Soil Water Conserv Res. 2013;1(2):53–63. https://doi.org/10.1016/S2095-6339(15)30039-3.
25. Ridwansyah I, Apip A, Wibowo H, Rahmadya A, Susiwidiyaliza S, Handoko U, et al. The impact of land use and climate change on water and sediment yields in Batanghari Watershed, Indonesia. Sains Malays. 2023;52(3):705–21. https://doi.org/10.17576/jsm-2023-5203-03.

Chapter 6
Decision Support Tool 2: Fisheries, Habitat, and Flooded Rice Fields

Ivana Yuniarti, Setiya Tri Haryuni, Dimas A. Wibowo, Steven Solikhin, S. Wenceslaus Wewen Wira Permadi, and Nurhaliza Amalia Lestari

Abstract This chapter presents the decision support tool (DST) for simulating the impacts of selected management actions on the *ikan bilih* fishery. The selected management actions are unsustainable fishing efforts on the fish stock and the effects of water level regulation on the fish's habitat. Further, we also elaborate on the methods to determine no-catch zone areas and simulate water level regulations on the flooded rice field. We combined a hydroacoustic approach with maximum entropy models (MaxEnt) to estimate the absolute abundance and current standing stock. Then, we engaged the Schaefer Surplus Production Model (SSPM) to simulate the impacts of fishing gear regulations on the fish stock fluctuations. Again, MaxEnt model combined with ArcGIS 10.5 was used to measure the effects of water level regulations on the availability of the fish habitat. We overlaid the lake's topography, bathymetry, and baseline map in ArcGIS 10.5 to simulate the influence of water level management on the extent of the area of flooded rice fields. Based on the simulations, a water level of 362.5 above sea level and restrictions to 100 lift nets (bagan) are highly recommended, highlighting the importance of the continuation of the previously advised management responses like reduction of lift nets. Further, we

I. Yuniarti (✉)
Research Centre for Environmental Archaeology, Marine Archaeology, and Sustainable Culture, National Research and Innovation Agency (BRIN), Cibinong, Indonesia
e-mail: ivana.yuniarti@brin.go.id

S. T. Haryuni
Research Centre for Fisheries, National Research and Innovation Agency (BRIN), Cibinong, Indonesia

D. A. Wibowo
Directorate of Laboratory Management, Research Facilities, and Science and Technology Park, National Research and Innovation Agency (BRIN), Cibinong, Indonesia

S. Solikhin · N. A. Lestari
Department of Marine Science and Technology, IPB University, Bogor, Indonesia

S. Wenceslaus Wewen Wira Permadi
Department of Geophysics and Meteorology, IPB University, Bogor, Indonesia

K. Glenk et al. (eds.), *Decision Support Systems for Multisectoral Tropical Lake Management in Indonesia*, https://doi.org/10.1007/978-3-032-11268-2_6

propose several areas as no-catch zone areas, particularly on the eastern area of the lake, where natural barriers occur.

Keywords Fish · ikan bilih · Lake · Lake Singkarak · Lift net (bagan) · Fish stock · Water level · Habitat model · MaxEnt · Habitat

6.1 Chapter Introduction

This chapter presents the decision support tool (DST) for simulating the impacts of selected management actions on the *ikan bilih* fishery. The selected management actions are (i) unsustainable fishing efforts on the fish stock and its supporting research, such as estimated initial and future stock (Fig. 6.1a), and (ii) the effects of water level regulation and water quality changing on flooded rice field (Fig. 6.1b) and on the *ikan bilih*'s habitat (Fig. 6.1c). Furthermore, the chapter also describes the simulation to determine no-catch zone areas.

To select management actions for Lake Singkarak to be included in the DST, we first researched the impacts of habitat change and the effects of unsustainable fishing gear. We also evaluated the influence of water level management on the availability and extent of the *ikan bilih* habitat.

6.2 The Impacts of Management of Unsustainable Fishing Gear on the Fish Stock

First, we analyzed the effects of unsustainable fishing gear, particularly Bagan/lift nets and *alahan* (traps). We referred to the work of [1], to obtain the values of catch probability for 3/4, 5/8, and 1 inch mesh size.

Next, we engaged the Schaefer Surplus Production Model (SSPM) [2, 3]. Some of the predicted coefficients from the SSPM are then used to create a stock growth algorithm (SGA) to predict future fish stock. A detailed methodology for building the SSPM is described in [4]. Meanwhile, the SGA, following Russel's concept embedded with Verhulst intrinsic population growth algorithms by [5], is conceptualized as:

$$S_{t+1} = S_t + \left(R.S_t\left(1-\left(\frac{S_t}{K}\right)\right)\right) - (M.S_t) - (F.N_t)$$

Thus $$S_1 = S_0 + \left(R.S_0\left(1-\left(\frac{S_0}{K}\right)\right)\right) - (M.S_0) - (F.S_0),$$

$$S_2 = S_1 + \left(R.S_1\left(1-\left(\frac{S_1}{K}\right)\right)\right) - (M.S_1) - (F.S_1) \text{ ...etc.}$$

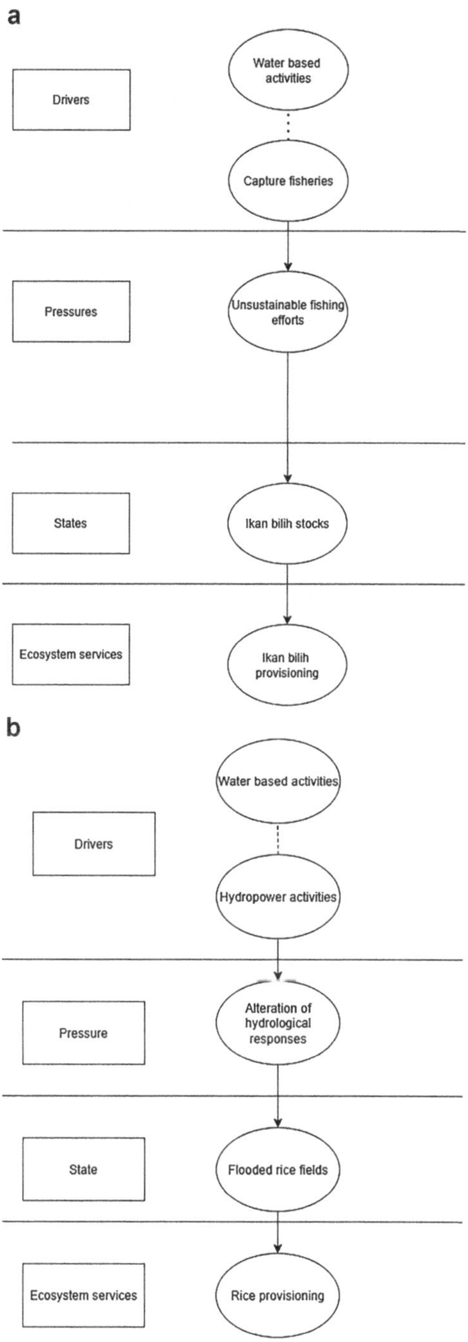

Fig. 6.1 The management actions considered in the DSS. (**a**) The impacts of unsustainable fishing efforts on the fish stock, (**b**) the influence of water level regulation on flooded rice fields, and (**c**) the effects of water level regulation and water quality changing on *ikan bilih* habitat. Initial stock prediction using fish absolute abundance approach

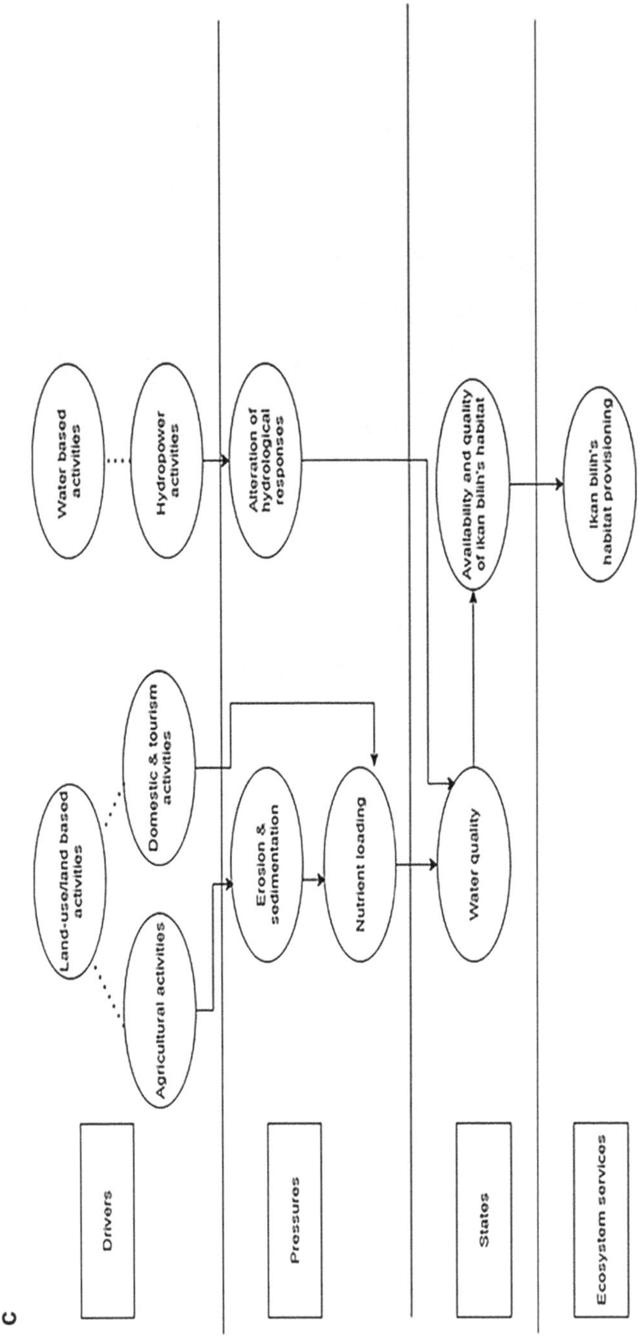

Fig. 6.1 (continued)

Note:

S_0 = The initial fish stock (Stock in 2023) = 900 and 2500 tons.
S_t = Fish stock in year t.
S_{t+1} = Fish stock in year t + 1.
R = Recruitment rate = 0.21 [6].
K = Carrying capacity = 2398.86 tons [7].
M = Mortality rate = 4.11 [6].
F=Fishing rate = 2.18 [8].
N_t = Number of fishing gears in year t.

The initial fish stock was estimated using a hydroacoustic approach (Sect. 6.1.1), while the carrying capacity value (K) was calculated from the fisheries productivity estimation using chlorophyll as a proxy method [7]. Several estimation models generated values that are much higher than those used here (e.g., [8]); however, when we cross-checked their estimated value with the actual situation, we concluded that the estimation values were overestimated. For example, the estimated optimal production in [8] is 805.15 tons per year, and the optimum effort is 644.03 units annually. At the same time, the actual count data in 2022 showed that there were 317 lift nets [9], and the stock was already collapsing. Thus, we chose to use more realistic values produced by our research. This approach ensures that our data is validated with actual situations to enhance the data's legitimacy.

6.3 Initial Stock Prediction Using Fish Absolute Abundance Approach

To aid stock prediction, we used data from the hydroacoustic equipment SIMRAD EK15 and length–weight relationships from enumerator-recorded data. First, we conducted laboratory experiments to determine the fish species' suitable target strength (TS). Then, using the obtained TS, we operated the equipment in a field survey in February 2023. The survey was divided into data collection for model construction and field validation. The data gained on the last day of the field survey was used as the validation data. The TS is calculated by [10]:

$$TS = (\sigma_{bs})$$

The equipment was run in a zig-zag pattern across shallow areas (at water level 362.5 m) to represent the hot spot and non-hotspot areas. Areas were selected based on findings of a Maximum Entropy Model (MaxEnt) [12], described in Sect. 6.2 below. The equipment is controlled to target single and schooling objects and to estimate fish abundance. Fish abundance was calculated following the algorithm given by [10]:

$$TS = SV - \rho$$

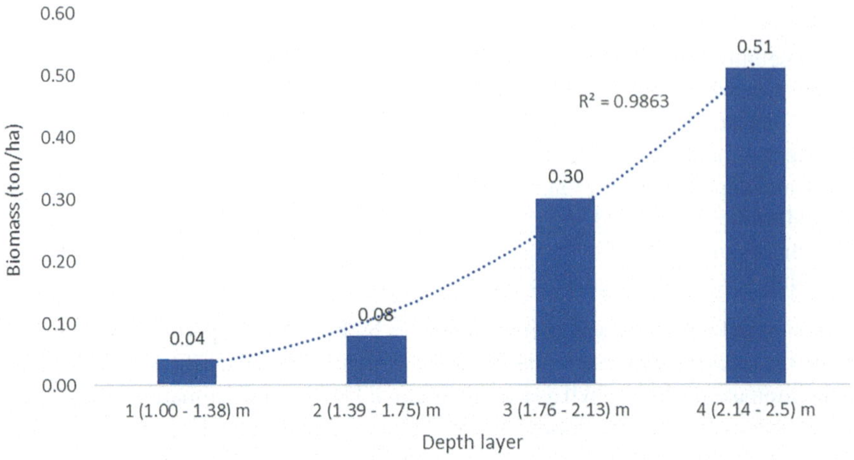

Fig. 6.2 Biomass of ikan bilih in four depth layers

Note:

TS = Target strength (dB).
σ_{bs}= Backscattering cross-section.
SV = Volume backscattering strength (dB).
ρ = Fish density per unit volume (ind/m^3).

The calculated fish abundance was divided into four depth categories, ranging from 0.04 to 0.51 ton/ha (Fig. 6.2). Although *ikan bilih* is famously known to inhabit shallow areas [11], the biomass is found to increase with depth (Fig. 6.2). Multiplying these results with the habitat extent obtained from the MaxEnt model (see Sect. 6.2 below), we calculated the initial fish stocks to lie in a range from 900 to 2500 tons.

6.4 The Effects of Water Level Regulation and Water Quality Changing on *Ikan Bilih* Habitat

We used Maximum Entropy Models (MaxEnt) [12] to simulate the influence of water level and fluctuating water quality parameters on the ikan bilih habitat. First, we modeled the water inundation level by combining topographic and bathymetric maps using ArcGIS 10.5. Next, we simulated the distribution of turbidity and temperature for each water inundation map. Turbidity is the selected water quality parameter to align with stakeholders' perspectives on the changing water quality caused by the hydropower company. Temperature is chosen because it can incorporate the effects of climate change on fish habitats.

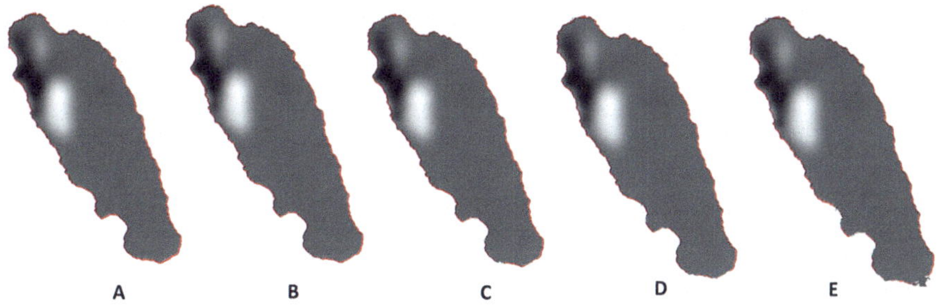

Fig. 6.3 Raster image of turbidity distribution for water level: (**a**) 361.5 m, (**b**) 362 m, (**c**) 362.5 m, (**d**) 363 m, and (**e**) 363.5 m

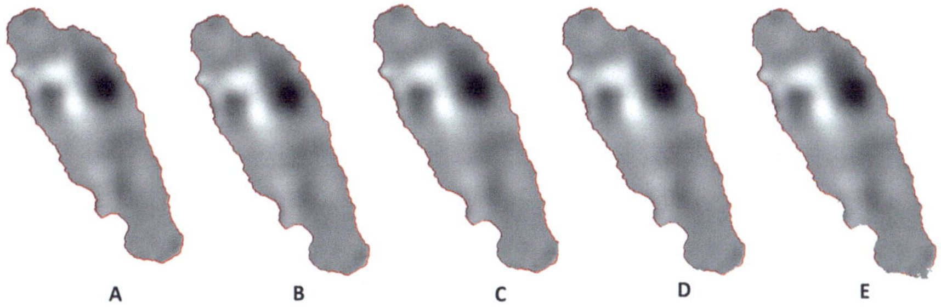

Fig. 6.4 Raster image of temperature distribution for water level: (**a**) 361.5 m, (**b**) 362 m, (**c**) 362.5 m, (**d**) 363 m, and (**e**) 363.5 m

We simulated every 50 cm water level increment from 360 to 370 m, resulting in 21 maps. Eventually, we selected five water levels to be presented in the DSS: 361.5 (the average factual water level obtained from five-year daily water level data), 362 m (the proposed water level by farmers), 362.5 (the minimum water level to support optimum electricity production), 363 m (the middle water level range to support optimum electricity production), and 363.5 m (the maximum water level appointed by the agreement between local customary group and the hydropower company –PLN Bukit Tinggi). Further detailed data collection and analysis methods are described in [4].

Data treatment to improve data credibility and incorporate uncertainties included collecting data series from 2022 to 2023 to represent both rainy and dry seasons, dividing data into training and test data, and engaging in cross- and field validation. Data legitimacy is maintained by ensuring data transparency and sharing. All data can be accessed at ivanayuniarti02/**Maxent-RF-catch**. The models reveal that changing the water inundation level does not significantly change turbidity and temperature distribution (Figs. 6.3 and 6.4). Meanwhile, the final simulation results concluded that regulating water levels above 362.5 m is desirable to maintain the availability of fish habitats. Below that level, the habitat extent drops significantly (Fig. 6.5).

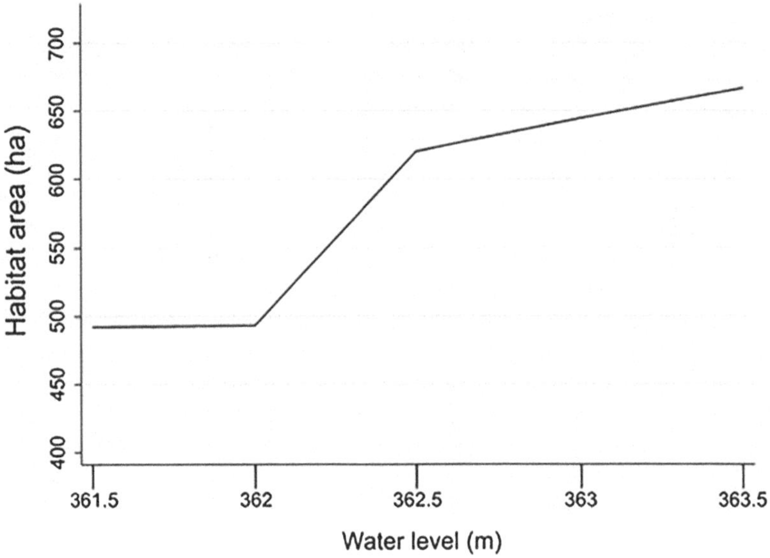

Fig. 6.5 Simulated *ikan bilih*'s habitat area for the selected water levels

6.5 Determination of No-Catch Zone Areas

The combination of water inundation and water quality distribution is then used as input for the MaxEnt models to select the most suitable areas for no-catch zones. No simulated significant difference in the zone pattern for each water level exists. Ultimately, we chose the 362.5 m water level area as the suggested area (Fig. 6.6).

The models simulated that the west and east parts of the lake are suitable no-catch zone candidates, marked by the red area in Fig. 6.6. However, the westside areas, such as *Nagari Sumpur* and *Paninggahan*, mostly have gradual slopes, making them accessible for traditional fishers. Thus, these *nagaries* are renowned as main *ikan bilih* habitats [11]. Considering this aspect, proposing these areas as no-catch zones may face greater opposition than proposing them for the eastern part, despite being in contrast to several suggestions by [13].

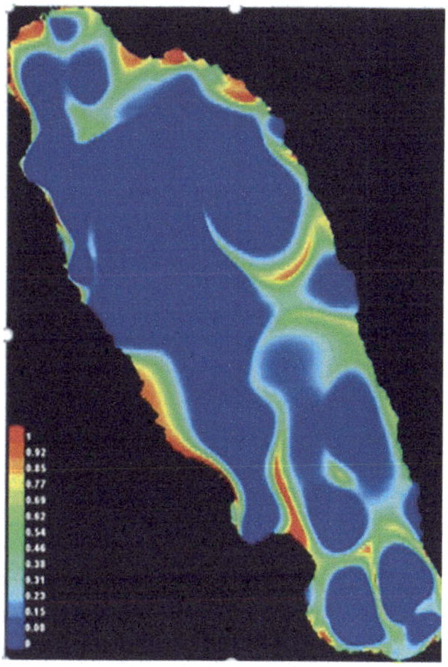

Fig. 6.6 Habitat of *ikan bilih* at 362.5 m water level. The red zones are the hotspots and might thus be considered as no catch-zone of the fish

6.6 The Impacts of Water Level Management on Flooding of Rice Fields

We overlaid the lake's topography, bathymetry, and baseline map in ArcGIS 10.5 to simulate the influence of water level management on the extent of the area of flooded rice fields. The overlaying processes include: (1) triangulated irregular network modeling [14], (2) raster modeling [15], (3) contour line interpolation [16], (4) overlaying stage [17].

The increased water level generally enlarges the extent of flooded areas, with a 0.5 m increase in water level elevating 1.4–14.4 ha of the flooded area (Fig. 6.7). Unfortunately, there are no recorded observation data to validate the results due to limited data series collection in Indonesia. However, the data generated from our approach is essential to calculate the lost economic value due to production loss (details in Chap. 8).

However, most of the rice fields are located next to the lake's shoreline, which commonly does not comply with formal regulations (e.g., [18]). Therefore, the results of this simulation cannot be directly applied to calculate requirements for a compensation scheme. To estimate such requirements, a coordination and consultation group should be established to obtain various perspectives, including the legal aspects.

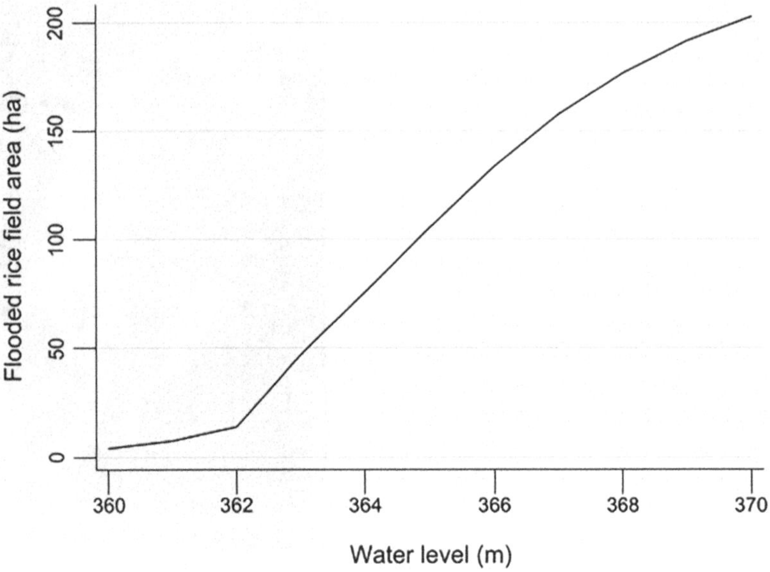

Fig. 6.7 The effect of water level management on the extent of the flooded rice fields

6.7 Conclusion

From our findings, we notice trade-offs between the habitat area of *ikan bilih* and the flooding of rice fields, which affects the economic value of the selected management actions. The trade-offs and other generated information, such as the predicted stocks, are the principal inputs for calculating the net present value and for undertaking cost-benefit analysis of alternative courses of management action (Chap. 8). Further, we highlight several sources of uncertainty in our models, such as the impact of rising temperatures due to global warming. Climate change affects almost all water uses: fish and habitat distributions, flooding, and fish recruitment; thus, it is essential to include climate change impacts in the models. Due to data, time, and funding limitations, the current DSS does not include this aspect. Thus, it becomes a principal target of future DSS development. Further elaboration regarding this topic can be read in Chap. 10.

References

1. Warsa A, Hendrawan ALS, Krismono. Penetapan ukuran mata jarin langli untuk penangkapan bilih (*Mystacoleucus padangensis*) di Danau Singkarak. J Fish Mar Res. 2020;4:178–86. https://doi.org/10.21776/ub.jfmr.2020.004.01.25.
2. Schaefer MB. Some aspects of the dynamics of populations important to the management of the commercial marine fisheries. Bull Math Biol. 1991;53:253–79. https://doi.org/10.1007/bf02464432.

3. Schaefer MB. Some considerations of population dynamics and economics. J Fish Res Board Canada. 1957;14:669–81. https://doi.org/10.1139/f57-025.
4. Yuniarti I, Wibowo DA, Triharyuni S, Suryono T, Anwar M. Fishing pressure or deteriorated habitat? Modeling approaches to identify the underpinning cause of hydropower and fishery conflict in an Indonesian tropical lake. Freshw Ecol. In Prog. 2024;
5. Vivekanandan E. Concept and objectives of stock assessment. Lecture Note Series No. 2/2017. 2017. Available: http://eprints.cmfri.org.in/12171/1/10-Concept%20and%20objectives%20of%20stock%20assessment.pdf.
6. Triharyuni S, Rahmadi P, Puspasari R, Rachmawati PF, Prianto E. Vulnerability of endemic bilih fish *Mystacoleucus padangensis* Bleeker, 1852 in Lake Singkarak, West Sumatra, Indonesia. IOP Conf Ser Earth Environ Sci. 2022;1119 https://doi.org/10.1088/1755-1315/1119/1/012010.
7. Triharyuni S, Rachmawati PF, Puspasari R, Nugroho D, Rahmadi P, Prianto E, et al. The development of capture fisheries in Lake Singkarak, West Sumatera, Indonesia: a study of carrying capacity. 2024.
8. Heri ER, Yulinda E, Arief H, Sosial J, Perikanan E, Perikanan F, et al. Analisis Bioekonomi Ikan Bilih (*Mystacoleuseus padangensis*) di Danau Singkarak Provinsi Sumatera Barat. J Sos Ekon Pesisir. 2022;3:12–22.
9. Fachri F, Hafil M. 317 Unit Bagan di Danau Singkarak Ancam Kelestarian Ikan Bilih. Pengoperasian bagan di Danau Singkarak mengancam ikan bilih. Republika online. 2022. Available: https://news.republika.co.id/berita/rlur0z430/317-unit-bagan-di-danau-singkarak-ancam-kelestarian-ikan-bilih
10. Maclennan DN, Fernandes PG, Dalen J. A consistent approach to definitions and symbols in fisheries acoustics. ICES J Mar Sci. 2002;59:365–9. https://doi.org/10.1006/jmsc.2001.1158.
11. Syandri H, Aryani N, Azrita. Karakteristik populasi dan habitat pemijahan ikan bilih (*Mystacoleucus padangensis* Blkr) di Danau Singkarak, Sumatera Barat. In: Seminar Nasional Fakultas Perikanan dan Ilmu Kelautan Universitas Riau. Pekanbaru: Fakultas Perikanan dan Ilmu Kelautan Universitas Riau; 2011. p. 94–9.
12. Phillips SJ, Anderson RP, Schapire RE. Maximum entropy modeling of species geographic distributions. Ecol Model. 2006;190:231–59. https://doi.org/10.1016/j.ecolmodel.2005.03.026.
13. Warsa A, Astuti DWHTLP. Upaya Konservasi Ikan Bilih (*Mystacoleucus padangensis*) dengan Penetapan Ukuran Tangkap dan Suaka di Danau Singkarak. Airaha. 2020;IX:036–48.
14. Arfaini J, Handayani HH. Analisa Data Foto Udara untuk DEM dengan Metode TIN, IDW, dan Kriging. J Tek ITS. 2016;5 https://doi.org/10.12962/j23373539.v5i2.17382.
15. Abdulghani T, Ubaedilah E. Analisis Dan Perancangan Sistem Informasi Geografis Sebaran Tanah Penduduk. Produktif J Ilm Pendidik Teknol Inf. 2018;2:1–12. Available: https://journal.umtas.ac.id/index.php/produktif/article/view/211
16. Afani IYN, Yuwono BD, Nurhadi B. Optimalisasi Pembuatan Peta Kontur Skala Besar Menggunakan Kombinasi Data Pengukuran Terestris dan Foto Udara Format Kecil. J Geod Undip. 2019;8:180–9.
17. Rachmah Z, Rengkung MM, Lahamendu V. Kesesuaian Lahan Permukiman di Kawasan Kaki Gunung Dua Sudara. J Spasial. 2018;5:118–29.
18. Permen Pupr. Peraturan Menteri Pekerjaan Umum Dan Perumahan Rakyat Republik Indonesia Nomor 28/PRT/M/2015 Tentang Penetapan Garis Sempadan Sungai Dan Garis Sempadan Danau. 2015 p. 6. Available: https://peraturan.bpk.go.id/Details/159992/permen-pupr-no-28prtm2015-tahun-2015

Chapter 7
Decision Support Tool 3: Tourism and Environment

Heri Apriyanto, Hermawan Prasetya, Sri Handoyo Mukti, Warseno, Aphang Suhendra, Tukiyat, and Taufiq Dwi Tamtomo

Abstract This chapter emphasizes the construction of a decision support tool (DST) in the DSS, elucidating the relationship between tourism and water quality using system dynamics (SD) modeling. We divide the environmental sub-model into three categories: tourism, water quantity, and water quality, also applied several important assumptions in building the models. Our Stock-Flow Diagram (SFD) model provides a detailed analysis of the complex interactions within the tourism industry at Lake Singkarak, highlighting the interconnected relationship between tourists, tourism revenue, and environmental sustainability. The model's scope is limited to analyzing annual changes, encompassing the period from 2013 to 2022 for existing conditions and projecting future scenarios from 2023 to 2042. The water quantity sub-model comprehensively analyzes rainfall dynamics, inflow and outflow rates, lake volume and elevation changes, and hydroelectric power generation. The water quality sub-model focuses on the pollutants entering the lake through rivers, fish farming activities, and tourism. We found that the current number has not significantly affected the quality of the lake's water, unlike other anthropogenic activities, such as agriculture. We highlighted that increasing the number of tourists up to threefold is still possible, assuming that all the pre-requisite assumptions are met. Overall, our current models are sufficient to inform sustainable tourism man-

H. Apriyanto (✉) · S. H. Mukti · A. Suhendra · T. D. Tamtomo
Research Centre for Sustainable Industrial and Manufacturing Systems, National Research and Innovation Agency (BRIN), Jakarta, Indonesia
e-mail: heri014@brin.go.id

Warseno
Directorate of Political, Legal, Defense, and Security Policies, National Research and Inovation Agency (BRIN), South Tangerang, Indonesia

H. Prasetya
Directorate of Economy, Employment, and Regional Development Policy, National Research and Innovation Agency (BRIN), Jakarta, Indonesia

Tukiyat
Research Centre for Limnology and Water Resources, National Research and Innovation Agency (BRIN), Cibinong, Indonesia

K. Glenk et al. (eds.), *Decision Support Systems for Multisectoral Tropical Lake Management in Indonesia*, https://doi.org/10.1007/978-3-032-11268-2_7

agement strategies for the lake. Initiatives such as expanding wastewater treatment facilities and promoting eco-friendly tourism practices can enhance economic benefits while protecting the lake's ecosystem.

Keywords Tourism · Lake · Lake Singkarak · Water quantity · Water quality · Visitors · System dynamics · Sustainability

7.1 Chapter Introduction

The lake is one of the top tourist destinations in West Sumatra, holding strategic value in supporting regional economic development, particularly in Solok and Tanah Datar Districts. As the second-largest lake on Sumatra Island, Lake Singkarak offers captivating natural beauty and is a hub for cultural and ecotourism activities. Various annual events, such as the cycling event Tour de Singkarak, significantly enhance the quality of tourism infrastructure in West Sumatra, catalyzing sports tourism development while benefiting local communities and the region in the long term [1].

This chapter emphasizes the construction of a decision support tool (DST) in the DSS, elucidating the relationship between tourism and water quality using system dynamics (SD) modeling (Fig. 7.1). SD can be seen as a method for modeling and simulating a complex system over a specific period, comprising problem understanding and system conceptualization, evaluation, and scenario development (Fig. 7.2). The complex system nature is reflected through relationships between variables, which sometimes have a feedback structure so that they are nonlinear and change over time [2]. SD utilizes feedback control and dynamic modeling to simulate complex systems, allowing for analysis of how various components interact over time through feedback loops and delays presented as a causal loop diagram (CLD) [3–5]. SD is particularly effective in understanding the behavior of dynamic systems, such as water resource management, where it helps visualize the interconnections and feedback processes that influence system performance [6].

In our application, the SD model addresses various factors influencing the condition of Lake Singkarak, including water quality and pollutant load, and connects these factors for the growth of tourism in the area. The tourism sub-model forecasts the future tourist numbers and their impact on the local and regional economy. By adopting this approach, we can enhance the integration and comprehensiveness of the developed SD model.

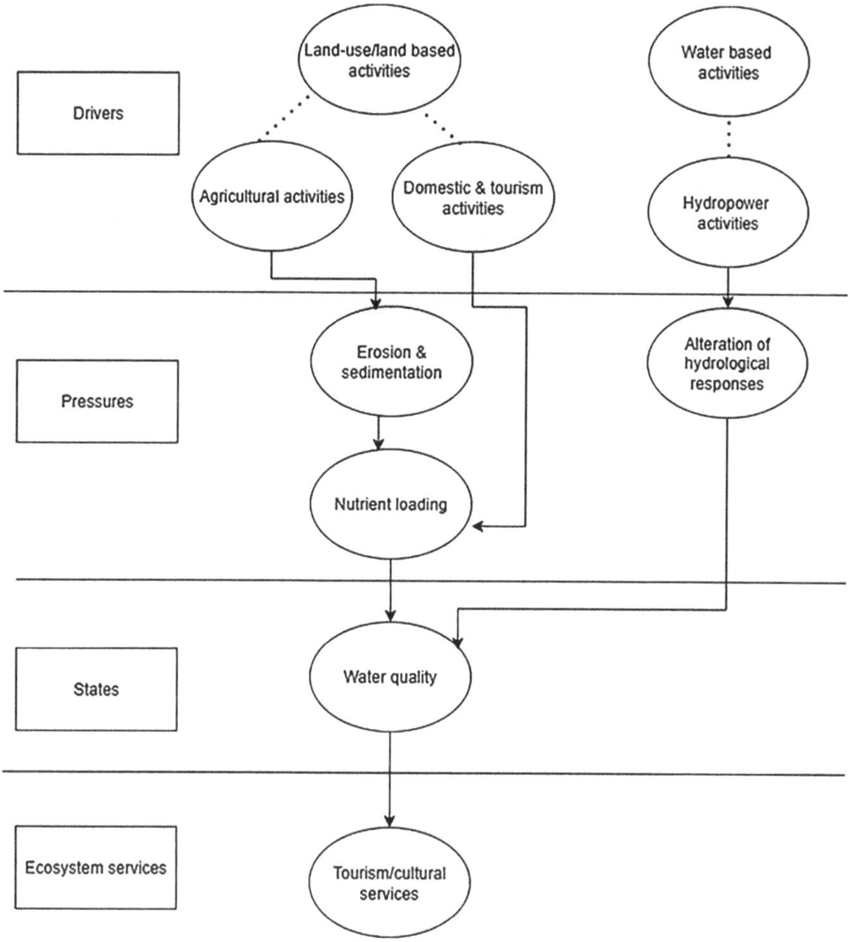

Fig. 7.1 The contribution of tourism and the water quality decision support tool (DST) in the DSS

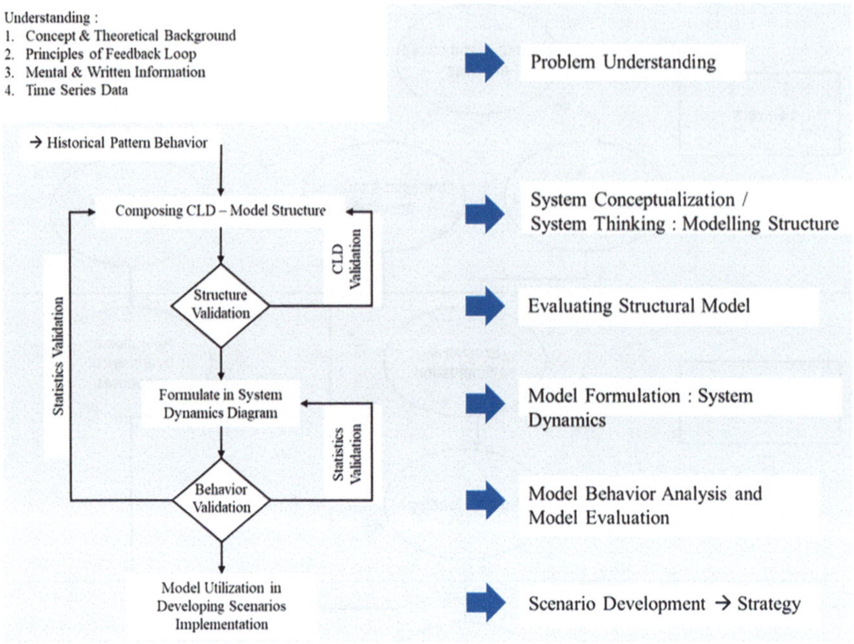

Fig. 7.2 System dynamics modeling steps

7.2 Tourism and Environmental Model of Lake Singkarak

We divide the environmental sub-model into three categories: (1) tourism, (2) water quantity, and (3) water quality. We collected field data and observations in the local culinary areas surrounding the lake embankments and collected secondary data from local institutions to validate the models. It is important to note that the water quality referred to in this chapter is the average measurement value across the lake, the inflow is associated with the aggregate of water flowing to the lake through all inlets, and the outflow is a combination of the discharge of the hydropower intake and the natural outlet at Ombilin Weirs.

The assumptions related to the tourism model include:

(a) The growth of domestic tourism is equivalent to West Sumatra's economic growth rate. Meanwhile, the growth of international tourism is projected based on the historical growth rate of foreign visitors. These assumptions align tourism projections with broader economic and sectoral trends.
(b) The availability of accommodation, specifically the total number of rooms, constrains the increase in the number of tourists. This reflects the practical limitation that accommodation capacity is a limiting factor for tourism growth.
(c) Tourism events can potentially increase the number of visitors by up to 100%.

7.2.1 Tourism Sub-model

A complex network of feedback loops, both positive (presented as blue arrows in Fig. 7.3) and negative (black arrows), shapes the tourism dynamics surrounding Lake Singkarak (as shown in Fig. 7.3). Positive feedback mechanisms, or reinforcing loops, are crucial for increasing tourist numbers to the area. An example of a reinforcing feedback loop is increased tourism revenue, which enables more significant investment in infrastructure such as roads, accommodations, and recreational facilities. This enhanced infrastructure not only improves accessibility but also enhances visitor satisfaction [7], leading to positive word-of-mouth recommendations and a subsequent rise in tourist arrivals and revenue. Furthermore, if successful, these marketing strategies create another reinforcing loop, as increased tourist interest leads to higher revenue that can be reinvested in additional promotional efforts, promoting a continuous growth cycle. Considering this, it is essential to anticipate overtourism that will occur, as has happened in Bali.

Additionally, the interconnected nature of environmental and community sustainability plays a crucial role in shaping the impacts of tourism. The growth in revenue generated by tourism can facilitate increased engagement from the local community in managing tourist activities, ultimately leading to better preservation of environmental resources, particularly in terms of water quality. This improved environmental condition, in turn, enhances the overall experience for visitors, who value pristine surroundings and are more inclined to revisit or endorse the destination [7], which eventually fosters a positive cycle of sustainable tourism

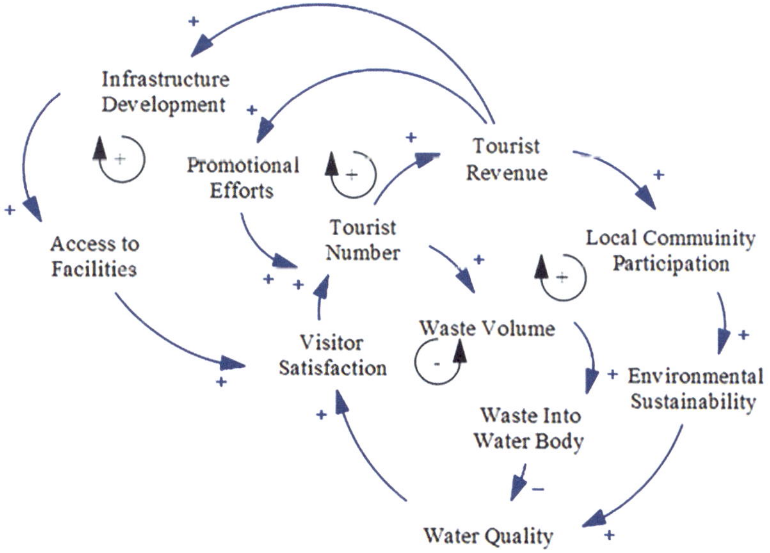

Fig. 7.3 Causal loop diagram (CLD) tourism at Lake Singkarak

development. Moreover, conserving cultural and ecological authenticity adds to a destination's appeal, drawing in environmentally conscious travelers.

On the other hand, balancing loops, also known as harmful feedback mechanisms, play an essential role in regulating tourism growth to prevent resource depletion. Increasing the number of tourists can lead to adverse environmental effects, such as decreased water quality and ecosystem damage. This decline in environmental sustainability can diminish the destination's attractiveness, resulting in lower visitor satisfaction and decreased tourist numbers. This feedback loop is a natural limitation, underscoring the importance of adopting sustainable practices that align tourism growth with environmental preservation. By grasping these dynamics, stakeholders can devise strategies that amplify positive feedback while mitigating the repercussions of negative feedback, ultimately ensuring the long-term sustainability and resilience of the tourism industry.

The Stock-Flow Diagram (SFD) model provides a detailed analysis of the complex interactions within the tourism industry at Lake Singkarak, highlighting the interconnected relationship between tourists, tourism revenue, and environmental sustainability (Fig. 7.4). Key components of this model include various stock variables: the total

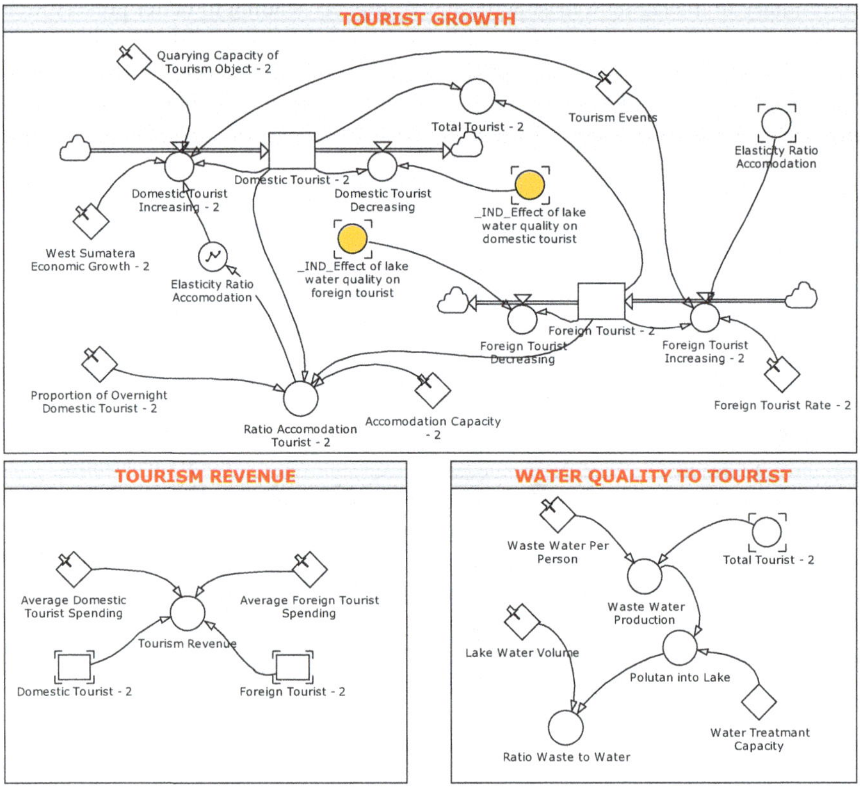

Fig. 7.4 Stock-flow diagram tourism at Lake Singkarak

count of domestic and international tourists as a measure of visitor traffic, tourism revenue as an indicator of economic impact from average expenditure, accommodation capacity for lodging availability, and the lake's water volume and pollution levels reflecting the environmental well-being of the region (as shown in Fig. 7.4).

The flow variables play a pivotal role in understanding tourism dynamics around Lake Singkarak. These variables include fluctuations in tourist numbers, influenced by factors such as water quality, accommodation capacity, and marketing efforts; wastewater generation per visitor based on total arrivals; pollutant entry into the lake from untreated wastewater; and tourism revenue, which is directly related to visitor numbers and their spending habits. The relationships among these variables are essential.

Water quality, determined by the waste-to-water volume ratio, significantly impacts tourist numbers, especially among international visitors, who are more sensitive to deterioration of environmental conditions. The amount of wastewater generated per tourist is linked to the capacity of water treatment facilities; inadequate treatment results in pollution, affecting water quality.

Additionally, accommodation capacity plays a role in shaping the visitor experience, with the availability of lodging influencing the destination's appeal. This model highlights the importance of feedback mechanisms. A positive feedback loop occurs when improved water quality—achieved through increased wastewater treatment capacity—attracts more tourists and increases tourism revenue. Conversely, a negative feedback loop occurs when a rise in tourist numbers surpasses the capacity of wastewater treatment facilities, leading to increased pollution and declining water quality, ultimately reducing tourist arrivals.

7.2.2 Interaction of Water Quantity and Water Quality Sub-model with Tourism Sub-model

This section aims to simulate various phenomena that affect the lake and its surrounding environment. One major issue is the utilization of lake water for electricity generation, which could result in the inundation of community land. Moreover, pollution from rivers and cage aquaculture enters the lake, leading to declining water quality. It is important to note that the number of cage aquaculture is still very limited; therefore, its influence is low. This deterioration in water quality deters tourists, reducing the lake's attractiveness. Conversely, increasing the number of tourists introduces additional pollutants into the lake, further degrading the water quality in a vicious cycle.

The model's scope is limited to analyzing annual changes, encompassing the period from 2013 to 2022 for existing conditions and projecting future scenarios from 2023 to 2042. The water quantity sub-model comprehensively analyzes rainfall dynamics, inflow and outflow rates, lake volume and elevation changes, and hydroelectric power generation. The water quality sub-model focuses on the pollutants entering the lake through rivers, fish farming activities, and tourism. It studies the behavior of these pollutants within the lake and their eventual outflow. The specific pollutants analyzed include Total Suspended Solids (TSS), Dissolved Oxygen (DO), Chemical Oxygen Demand (COD), Biological Oxygen Demand (BOD), nitrite, nitrate, Total Phosphorus (Total P), and Total Nitrogen (Total N).

Additionally, the model investigates the dynamics of tourist visits to the lake, providing insights into how water quality and quantity changes influence tourism. It explores the interconnected relationships between water quantity, water quality, and tourist numbers, offering a holistic view of how these factors interact and impact the lake's ecosystem. The following CLD (Fig. 7.5) maps the model that illustrates the interconnections between the dynamics of lake water quantity, lake water quality, and the impact of tourism.

The water quantity sub-model describes how much water enters and leaves the lake and how liquid waste is brought in and out. This sub-model also explains electricity generation and the resulting inundation. The impact is an increase in the amount of liquid waste and inundation created (as shown in Fig. 7.6).

The sub-model of water quality is constructed based on the changes in the stock of pollutants in the lake, which are influenced by the inflow that carries

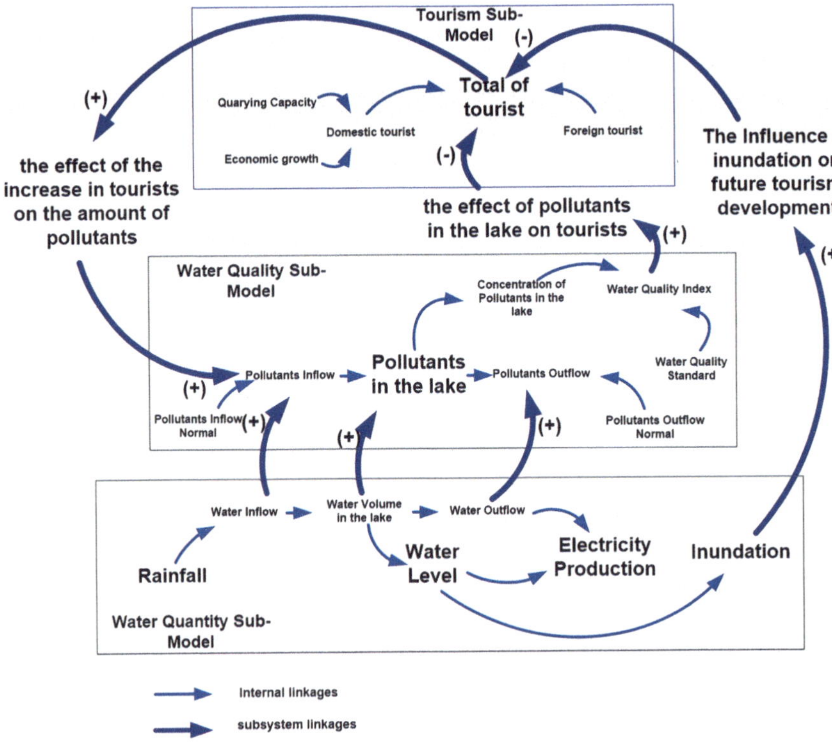

Fig. 7.5 Causal loop diagram (CLD) of the relationship between lake water quantity and quality and tourism

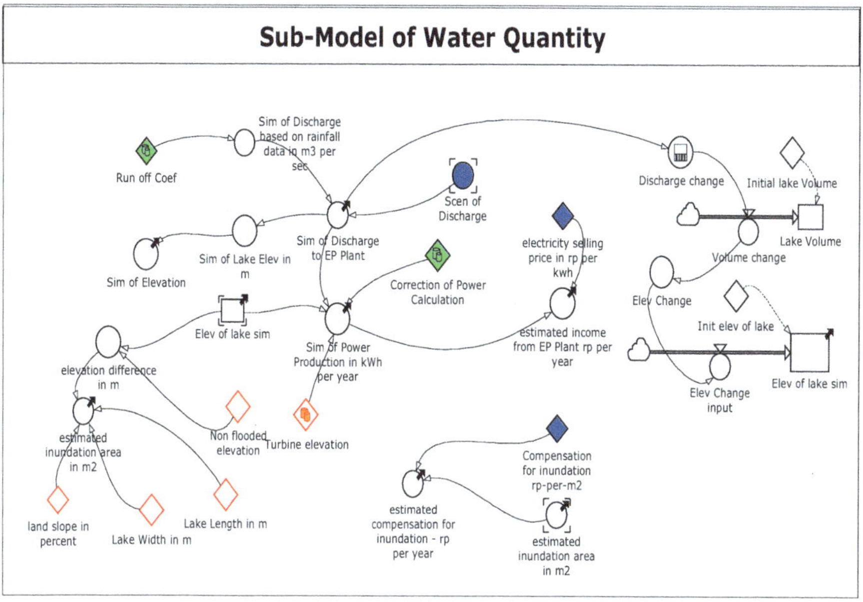

Fig. 7.6 Stock-flow diagram water quantity

contaminants and the outflow through which some pollutants are removed. The difference between the inflow and outflow of pollutants results in the accumulation of contaminants in the lake water. As this model operates annually, detailed natural phenomena such as dispersion, absorption, and diffusion are not discussed. The values or concentrations of pollutants obtained are annual averages (as shown in Fig. 7.7).

7.2.3 *Validation*

7.2.3.1 Water Quantity Sub-model Validation

Validation is carried out on the model structure and its numerical output based on statistical data. The model structure is validated by discussions with parties responsible for water quality, resulting in the final CLD model. Meanwhile, the Mean Absolute Percentage Error (MAPE) method numerically validates the model output results [8].

$$MAPE = \frac{1}{n}\sum_{t=1}^{n}\left[\frac{Xm - Xd}{Xd}\right]x\,100\% \tag{7.1}$$

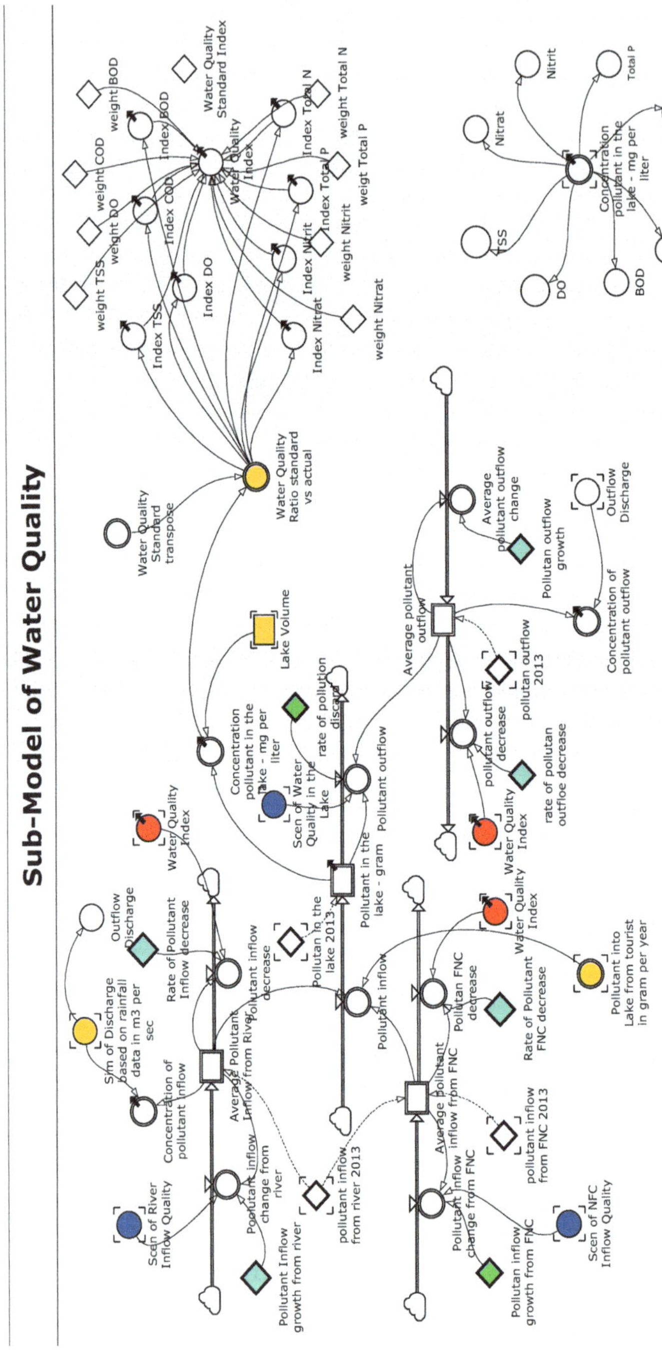

Fig. 7.7 Stock-flow diagram water quality

where

MAPE: the Mean Absolute Percentage Error
Xd = Data
Xm = Result of simulation
n = Number of data

If the MAPE value is <5%, the model is considered very accurate; if it is between 5% and 10%, it is considered accurate; and if the MAPE value is >10%, it is considered less precise.

The simulation results are summarized in Table 7.1, elucidating that the model is highly accurate. Overall, the validation result for each parameter shows very accurate results except for water quality in the lake's outflow, which generates accurate result.

(A) **Flow Rate Validation Results**

The simulation results were calculated based on rainfall data and runoff coefficients. This involved comparing flow rate data from 2013 to 2022 with calculated flow rates derived from rainfall. The resulting comparison is illustrated in Fig. 7.8.

Table 7.1 Summary of model validation results

No	Sub-model	Parameter	MAPE value (%)	Conclusion
1	Water quantity	Flow rate	3.77	Very accurate
		Lake elevation	0.26	Very accurate
		Electricity production	4.37	Very accurate
2	Water quality	Lake water quality	3.41	Very accurate
		Lake inflow quality	4.53	Very accurate
		Lake outflow quality	5.23	Accurate

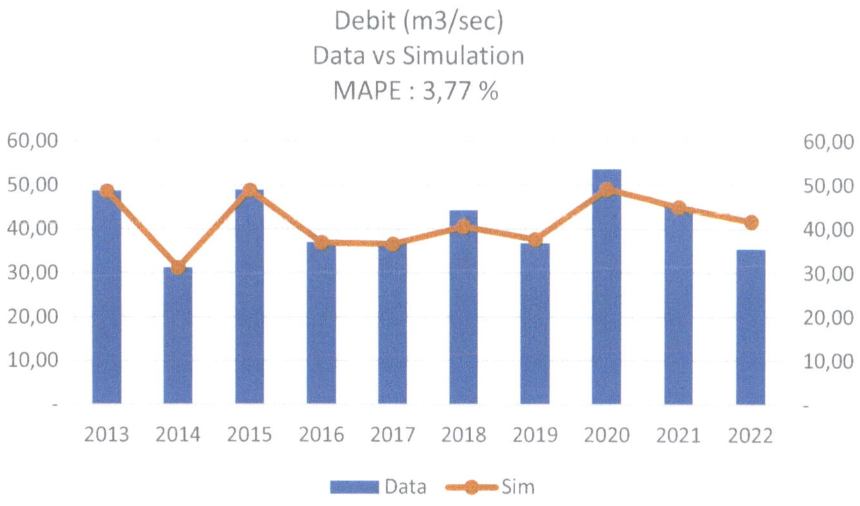

Fig. 7.8 Flow rate data versus simulation results

The simulation of flow rate calculations based on rainfall data from 2013 to 2022 resulted in a MAPE score of 3.77%, which indicates that the simulation results can be considered very accurate (Table 7.1).

(B) **Water elevation validation results**

The validation results of water elevation also show a high level of accuracy, presented by a 0.26% MAPE score (Fig. 7.9 and Table 7.1).

7.2.3.2 Water Quality Sub-Model Validation

(A) **Validation results of pollutant concentration in the lake's waters**

The simulation was conducted by calculating the changes in water quality from inflows through rivers and cage aquaculture, as well as outflows from the lake, between 2013 and 2022. Validation was performed by comparing the simulation results of lake pollutant concentrations in 2022 with actual water quality measurement data collected in 2022 [9]. Figure 7.10 illustrates the comparison between the lake pollutant concentration data and the simulation results for 2022. This comparison obtained a MAPE score of 3.41% and shows that the simulation results are very accurate (Table 7.1).

(B) **Validation results of pollutant concentration in the lake's inflow**

The simulation was conducted by calculating changes in water quality entering through rivers (average) and cage aquaculture from 2013 to 2022. Validation was performed by comparing the simulation results of pollutant concentrations in the lake inflow (rivers and cage aquaculture) with actual water quality measurement

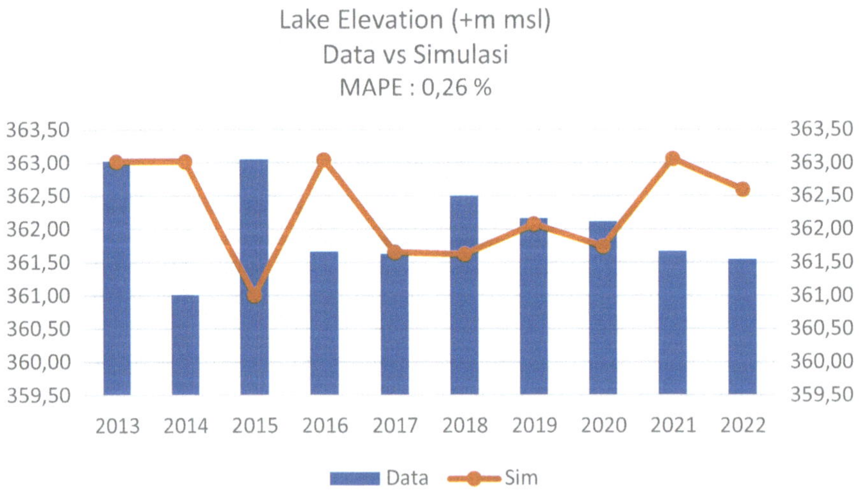

Fig. 7.9 Lake water elevation versus simulation results

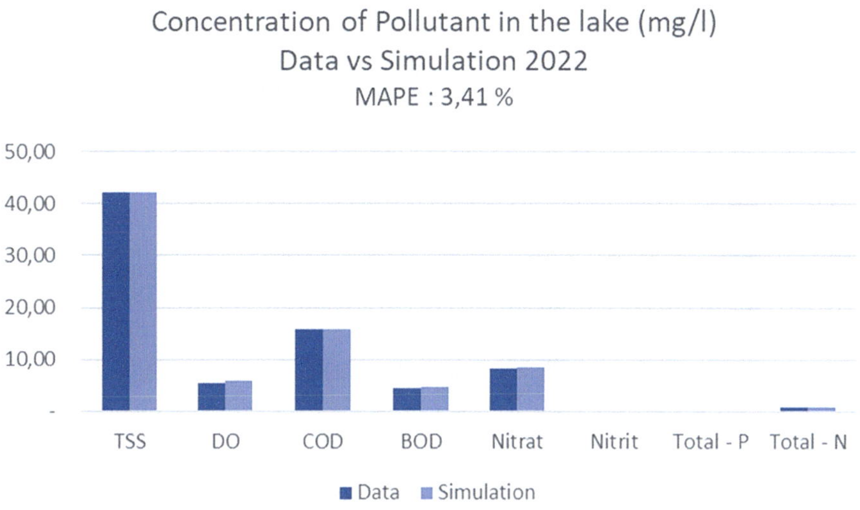

Fig. 7.10 Lake pollutant concentration versus simulation result

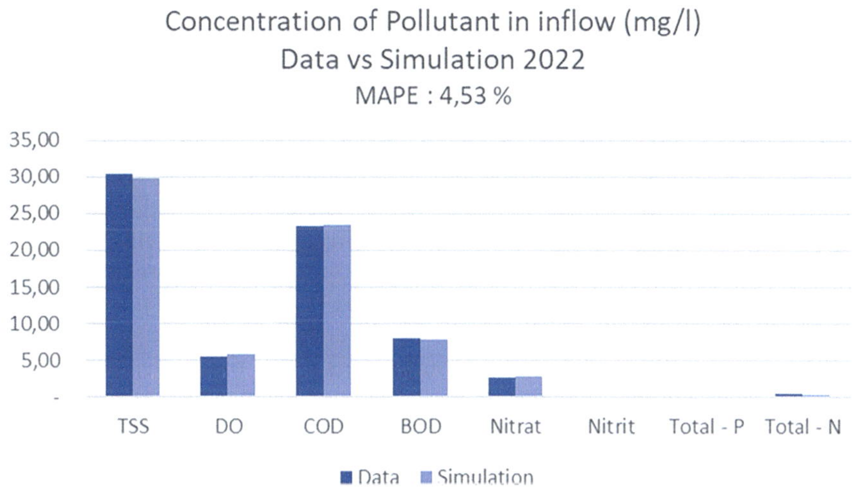

Fig. 7.11 Pollutant concentration in the lake's inflow versus simulation results

data collected in 2022 [9]. A MAPE score of 4.53% was obtained, showing that the simulation results are very accurate (Fig. 7.11 and Table 7.1).

(C) **Validation results of pollutant concentration in lake outflow**

The simulation was conducted by calculating the changes in water quality exiting the lake through the Weir Ombilin and the Lake Singkarak hydropower company (PLTA) intake from 2013 to 2022. Validation was performed by comparing the

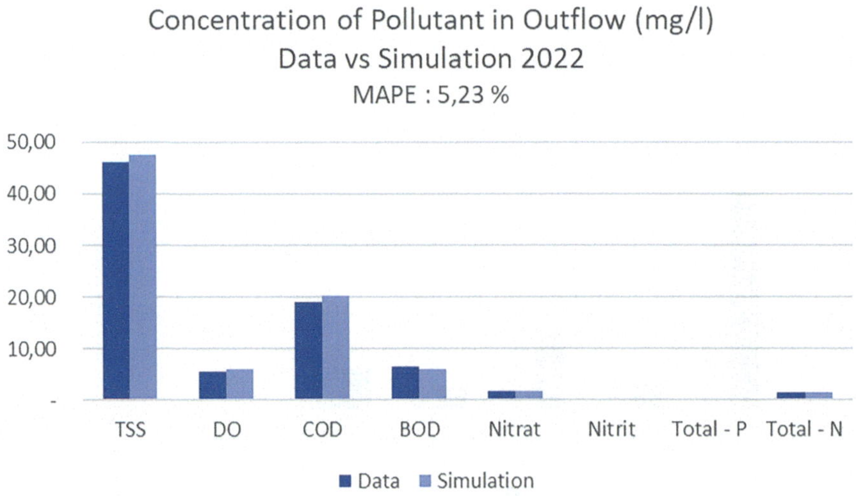

Fig. 7.12 Pollutant concentration outflow versus simulation results

simulation results of pollutant concentrations at the lake outflow with actual water quality measurement data collected in 2022. Fig. 7.12 compares the pollutant concentration data at the lake outflow and the simulation results for 2022. From the comparison between the actual data and the simulation results of pollutant concentrations in the lake outflow, a MAPE score of 5.23% was obtained. This indicates the simulation results can be considered to be accurate, reflecting reliable outcomes (Table 7.1).

- **Simulations of Tourism and Lake Sustainability: Impact Prediction**

According to the CLD in Fig. 7.5, the increase in the number of tourists will affect the pollutants entering the lake, thereby decreasing lake water quality. On the other hand, if water quality decreases, the number of tourists will decrease.

The current decline in lake water quality occurs due to river water entering the lake and anthropogenic activities on the lake's watershed and water bodies. The simulation results, with 10% and 20% of pollutants, will change lake water quality (Fig. 7.13).

The above results show that the water quality will be 2–5 times worse compared to the baseline due to increased pollutants entering the lake. As shown in the simulation results in Fig. 7.14, this will decrease the number of tourists by 10–15% relative to baseline estimates.

Based on the simulation results, increasing tourist attractions by 2–3 times enhances the number of tourists (Fig. 7.15). Interestingly, our simulation results on the impact of the increase in the number of tourists up to two to threefold show that the pollutant load is still under the current baseline data (Fig. 7.16). It implies that, under current assumptions, tourism activities do not produce a significant pollutant load, unlike agriculture and domestic activities. It can be understood that the present tourism activities, which are sightseeing, riding on banana boats, and dining, have

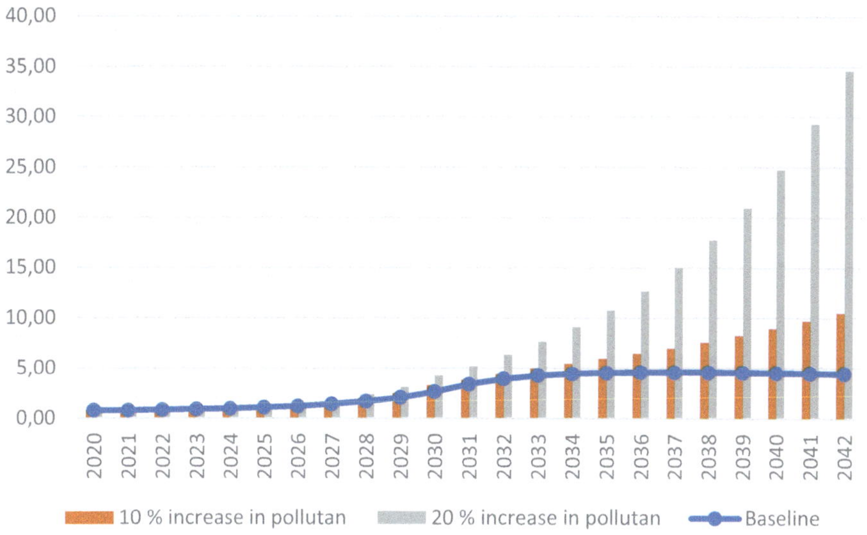

Fig. 7.13 The effect of increased pollutants on the water quality index

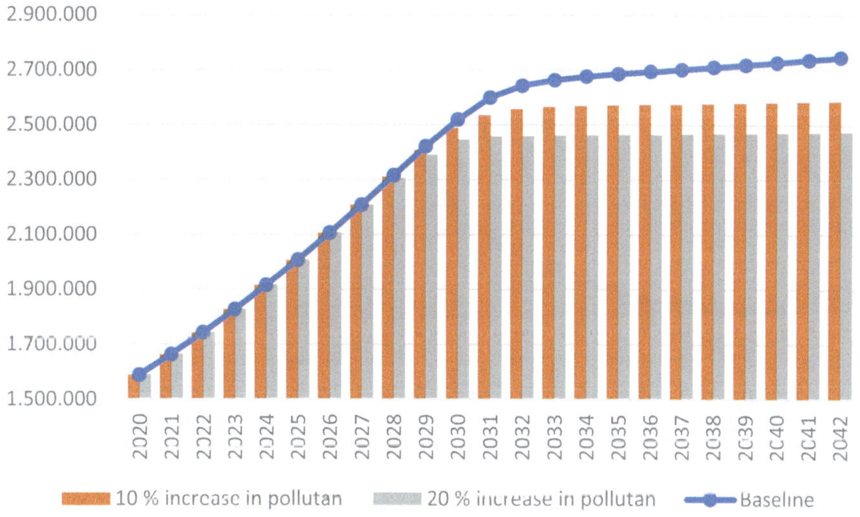

Fig. 7.14 The effect of water quality on the number of tourists

quite a low impact on the increase of organic pollutants. Thus, tourist numbers can still be endorsed, assuming that the current activities continue, and other factors are ceteris paribus. However, further assessment of total lake carrying capacity related to other water uses is required as a safety measure.

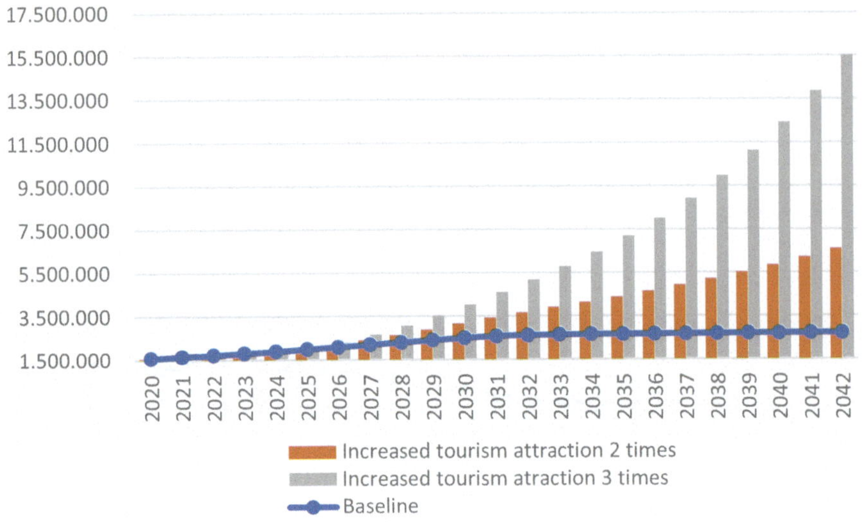

Fig. 7.15 Simulation results of the effects of increasing tourist attractions on the number of tourists

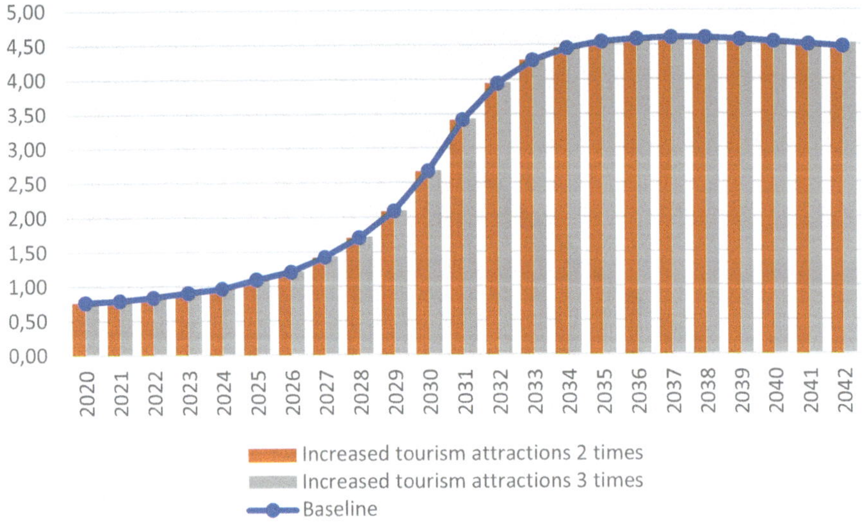

Fig. 7.16 The effect of increasing tourist attractions on the water quality index

7.3 Conclusion

Our models enable us to simulate the impact that will occur due to the decline in lake water quality on the number of tourists, as well as to predict changes in lake water quality if tourism activities, especially lake tourism, are increased. We found that the current number has not significantly affected the quality of the lake's water, unlike other anthropogenic activities, such as agriculture. We highlighted that increasing the number of tourists up to threefold is still possible, as it may not significantly cause a decline in water quality, assuming that all the pre-requisite assumptions are met. Further, we also notice that any decrease in the quality of the lake's water will decrease the interest or number of tourists. However, considering that our model validation runs in the lake inflow and outflow, it will be imperative to conduct further simulations and validation in the lake's essential tourism spots to obtain more conclusive evidence. We also underline the importance of the lake's total carrying capacity assessment as a safety measure.

Overall, our current models are sufficient to inform sustainable tourism management strategies for Lake Singkarak. Initiatives such as expanding wastewater treatment facilities and promoting eco-friendly tourism practices can enhance economic benefits while protecting the lake's ecosystem. Thus, we also recommend the SFD framework as a valuable analytical tool for exploring different scenarios and developing strategies that balance economic growth with environmental preservation in the region.

References

1. Susanti R, Priyambodo TK, Damanik J, Soeprihanto J. Organization of sport tourism event Tour De Singkarak contribution toward tourist attraction improvement in West Sumatera. IOSR J Human Soc Sci (IOSR-JHSS). 2017;22(10):1–13. https://doi.org/10.9790/0837-2210050113.
2. Aryawati NPA, Negara ISK. COVID-19: Kebijakan Strategis Dalam Menangani Dampak Sosial Ekonomi Dari Sisi Biaya Tidak Langsung di Indonesia. Sophia Dharma. 2021;3(1):1–11.
3. Achhami A, Kalra A, Ahmad S. Dynamic simulation of lake mead water levels in response to climate change and varying demands. World Environmental and Water Resources Congress 2018: Watershed Management, Irrigation and Drainage, and Water Resources Planning and Management—Selected Papers from the World Environmental and Water Resources Congress 2018. 2018:pp 260–8. https://doi.org/10.1061/9780784481400.023.
4. Ahmad S, Simonovic SP. System dynamics modeling of reservoir operations for flood management. J Comput Civ Eng. 2000;14(3):190–8. https://doi.org/10.1061/(ASCE)0887-3801(2000)14:3(190).
5. Apriyanto H, Prasetya H, Warseno, Suhendra A, Mukti SH, Kusharsanto ZS, et al. Linkage analysis of factors influencing the sustainability of Lake Singkarak in West Sumatra Province, Indonesia. IOP Conf Ser Earth Environ Sci. 2023;1267(1):012025. https://doi.org/10.1088/1755-1315/1267/1/012025.
6. Bozorg-Haddad O, Dehghan P, Zareie S, Loáiciga HA. System dynamics applied to water management in lakes*. Irrig Drain. 2020;69(4):956–66. https://doi.org/10.1002/ird.2470.

7. Yunitasari R. Analisis persepsi kondisi pariwisata dan valuasi ekonomi manajemen penanganan sampah makroplastik Danau Singkarak. Unpublished thesis. IPB University; 2024. Available http://repository.ipb.ac.id/handle/123456789/157078.
8. Al-Nuaami W, Dawod L, Kibria B, Ghorbani S. Design and implementation of a deep learning model and stochastic model for the forecasting of the monthly Lake water level. Limnolog Rev. 2024;24(3):217–34. https://doi.org/10.3390/limnolrev24030013.
9. Kementerian Pekerjaan Umum dan Perumahan Rakyat. Masterplan Revitalisasi Danau Singkarak 2022.

Chapter 8
Decision Support Tool 4: The Integrating Socio-economic Models

Ivana Yuniarti, Guruh Satria Ajie, Ragil Yunitasari, Rosetyati Utami, Djamhuriyah S. Said, Gunawan Pratama Yoga, and Bahroin Idris Tampubolon

Abstract This chapter elaborates on the monetization methods of the ecosystem services trade-offs related to the ecosystem-based management driver(s)-pressure(s)-state(s)-ecosystem services-response(s) (EBM-DPSER) framework and the calculation methods of the sustainability index (SI), which are the principal final outputs of the decision support system (DSS). The included parameters in the economic valuation are the influence of land-use best management practice on clean water provisioning, the impacts of water level management on *ikan bilih* habitat provisioning, the impacts of water level management on loss of rice production, no-catch zone establishment, and macroplastic management. We applied several economic valuation methods, such as market price approaches, habitat replacement cost, benefit transfers, and contingent valuation methods (CVM). We identified several uncertainties that need to be incorporated and underlined important updates for future research and DSS development. The results of the benefit-cost analysis are vital for the decision-making process in the area. Meanwhile, the SI, although newly introduced, can set a benchmark for future lake management.

Keywords Lake · Lake management · Economic valuation · Ecosystem services · Benefit-cost analysis · Sustainability · Sustainability index

---

I. Yuniarti (✉)
Research Centre for Environmental Archaeology, Maritime Archaeology, and Cultural Sustainability, National Research and Innovation Agency (BRIN), Cibinong, Indonesia
e-mail: ivana.yuniarti@brin.go.id

G. S. Ajie · R. Utami · D. S. Said · G. P. Yoga
Research Centre for Limnology and Water Resources, National Research and Innovation Agency (BRIN), Cibinong, Indonesia

R. Yunitasari · B. I. Tampubolon
Department of Resources and Environmental Economics, Faculty of Economy and Management, IPB University, Bogor, Indonesia

K. Glenk et al. (eds.), *Decision Support Systems for Multisectoral Tropical Lake Management in Indonesia*, https://doi.org/10.1007/978-3-032-11268-2_8

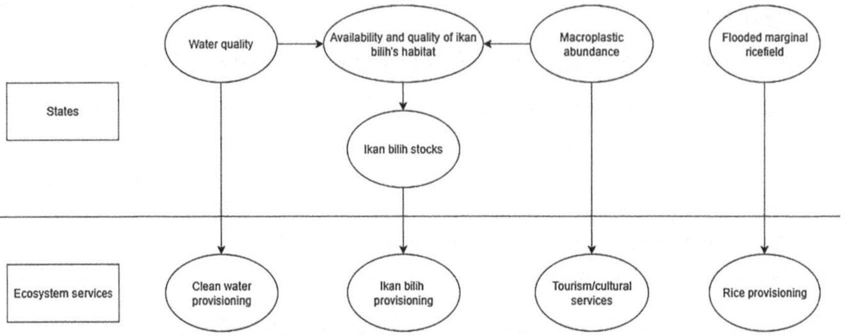

Fig. 8.1 The contribution of the economic valuation of ecosystem services in the DSS

8.1 Chapter Introduction

The economic valuation of ecosystem services (ES) is one of the main ways for science to find recognition in policy and decision-making [1]. This chapter quantifies the ES trade-offs related to the ecosystem-based management driver(s)-pressure(s)-state(s)-ecosystem services-response(s) (EBM-DPSER) framework (Fig. 8.1) within the decision support system (DSS). The chapter concludes with illustrations of some examples of ES trade-off analysis in the DSS's outputs.

We relate the economic valuation to the selected management scenarios (MS) (see Chap. 2), which include:

1. Watershed land-use best practice management scenarios
2. *Ikan bilih* fishing regulation
3. Water inundation level management
4. Establishment of no-catch zone
5. Macroplastic management

8.2 Economic Valuation

8.2.1 Influence of Land-Use Best Management Practice on Clean Water Provisioning

We selected the value transfer method due to the limited timeframe and funding for the project for assessing the impacts of turbidity increment on water treatment costs by referring to the meta-analysis work of [2]. Although the land-use best management practice (BMP) is associated with changes in several water quality parameters (Chap. 4), turbidity was selected as the key parameter due to its ease of monitoring and data collection by local government staff.

The value generated from the cited meta-analysis was selected as it was conducted in a Global South country in the tropics and subtropics, thus sharing similarity with our study area. The meta-analysis indicates that a 1% increase in turbidity causes a 0.1162% increment in water treatment costs. Based on this finding, and using population and treatment cost estimates relevant to the Lake Singkarak area, we calculate the economic value of improving the water quality through the best management practice as follows:

$$EV_{Tt} = \left(\frac{T_t - T_{t+1}}{T_t}\right) * \left(\frac{0.1162 * AVC}{100}\right) * \left(W * N\right) \tag{8.1}$$

where

EV_{Tt} = Economic value of turbidity changes in the year t
T_t = Average turbidity level in the year t obtained from the SWAT-QWET models
T_{t+1} = Average turbidity level in the year t + 1, the SWAT-QWET models
AVC = Average cost of water treatment (IDR/m^3)
W = Average local people's water consumption (m^3/person/year)
N = Number of people using Lake Singkarak's water (people)

We recognized that the economic valuation of land-use BMP on clean water provisioning should comprise changes in water quantity due to sedimentation rate alterations. However, due to data limitations and the project's short timeframe, this aspect was not included in the current update. More importantly, turbidity measurement can fluctuate highly due to various external factors such as weather conditions; thus, the input turbidity measurement data should incorporate uncertainties, which can be managed using several modeling tools such as Monte Carlo Simulation. Nevertheless, considering local government agencies may have limited human and equipment capacity to run such models, we utilized average turbidity measurement data in the DSS.

8.2.2 Economic Valuation of Ikan Bilih *Provisioning*

We applied the market price method to assess the ikan bilih provisioning by referring to the stock growth algorithm (SGA) (see Sect. 6.1). This valuation method, based on the farm price, is selected due to its nature to represent the willingness to pay for exchange rights. The price is the DSS user's input, corrected for inflation with the consumer price index (CPI).

The equation to calculate the economic value of *ikan bilih* provisioning is thus as follows:

$$EV_{st} = \left(S_t - S_{t-1}\right) * \left(P_t * \frac{CPI_t}{CPI_b}\right) \tag{8.2}$$

where

EV_{st} = Economic value of *ikan bilih* provisioning in the year t
S_t = Estimated fish stock in the year t (kg)
S_{t-1} = Estimated fish stock in the year t-1 (kg)
P_t = Average farm market price of *ikan bilih* in year t (IDR/kg)
CPI_t = Consumer price index in the year t
CPI_b = Consumer price index in the baseline year (2024)

The price is influenced by various sources of uncertainty [3]. Hence, this equation should be interpreted with consideration of the user input price data. Once again, our principal reason for using an average price in this algorithm is to accommodate the capacity limitations of local governmental staff in running uncertainty and risk models.

8.2.3 *Economic Valuation of the Impacts of Water Level Management on* **Ikan Bilih** *Habitat Provisioning*

There are two steps of valuing the impacts of water level management on *ikan bilih* habitat provisioning:

1. Determining changes in habitat extent: We selected a water level of 362.5 m as the baseline for habitat extent (see Sect. 6.2).
2. Valuing habitat extent changes: We applied a habitat replacement cost (HRC) approach to assign the economic value of the habitat extent alterations.

Considering that habitat extent significantly influences fish production, a production function approach would be ideal to assess its economic value. However, the absence of reliable fish catch records is a barrier to its application. Instead, we rely on our absolute fish abundance measurement data (see Chap. 6) and use market prices to assess their economic value adjusted with the CPI to incorporate inflation effects, following the formula:

$$EV_{Lt} = \left(\left(A_{Lt} * AD_{Lt}\right) - \left(A_{L362.5} * AD_{L362.5}\right)\right) * \left(P_t * \frac{CPI_t}{CPI_b}\right) \quad (8.3)$$

where

EV_{Lt} = Economic value of the effects of water level management on *ikan bilih* provisioning in the year t

A_{Lt} = The extent of habitat at water level management L in the year t obtained by the models elucidated in Sect. 6.2

$A_{L362.5}$ = The extent of habitat at water level management 362.5 in the year t obtained by the models elucidated in Sect. 6.2

AD_{Lt} = Average absolute fish abundance at water level L in the year t obtained by the models presented in Sect. 6.1

$AD_{L362.5}$ = Average absolute fish abundance at water level 362.5 in the year t obtained by the models presented in Sect. 6.1

P_t = Average farm market price of *ikan bilih* in year t (IDR/kg)

CPI_t = Consumer price index in the year t

CPI_b = Consumer price index in the baseline year (2024)

8.2.4 *Economic Valuation of the Impacts of Water Level Management on Loss of Rice Production*

To evaluate the effects of water level management, we first quantified changes in the extent of flooded rice field area, using the 362.5 m water level as the baseline information (details in Sect. 6.2). Secondly, we estimated the economic loss of rice production based on the compensation value paid by the hydropower company. This method likely generates lower economic value than other methods, such as contingent valuation and production function, as it is based on the agreement between the hydropower company and the farmers. However, we could not conduct a thorough production function approach due to the lack of labor data.

Inferring that the increment of water level from 362.5 m will increase the flooded area, the loss of rice production due to the reduction of harvested area is calculated as the economic impact of water level management. The value is defined as follows:

$$EV_{PRt} = ((A_{ft} - A_{362.5t} - A_{50t}) * C_t \tag{8.4}$$

where

EV_{Prt} = Economic value of the impacts of water level management on the loss of rice production (IDR)

A_{ft} = The extent of flooded area in year t (ha)

$A_{362.5t}$ = The extent of the flooded area at water level 362.5 m in year t (ha)

A_{50t} = The extent of the lake watershed, stated as the area within 50 m range from the lake beach line (ha). This area is not compensated for the flooded area according to the lake watershed regulation [4]

C_t = Compensation price (IDR/ha)

We noticed that various external factors, such as rainfall, can affect land productivity and consequently, the compensation price. However, until the time of writing, the agreed compensation rate remains at 1000 IDR/m^2 or 10,000,000 IDR/ha. While further analysis using production function methods or other alternative approaches is recommended, data limitations—particularly on labor—may impede the application of such methods in the study area.

8.2.5 Economic Valuation of No-Catch Zone Establishment

A habitat replacement cost (HRC) approach was selected to assess the economic value of the no-catch zone, following the guidelines by [5]. This method was preferred over the recommended production function method due to limitations in catch data and records of the fishing area extent.

The cost of semi-ex situ habitat construction was used as the initial economic value, and the annual operation and maintenance cost were used as the annual economic value of the no-catch zone. The data obtained from [6] indicated that habitat construction costs range from 1 to 3 million IDR per m^2, with annual operational and maintenance costs of 40 thousand IDR per m^2. These values serve as input for the benefit value in the DSS.

We distinguished the HRC based on the DSS user's selected no-catch zone area proximity to *ikan bilih* hotspots (see Sect. 6.4). We assumed that the areas within hotspots generate more economic value than those outside. Accordingly, the year-one economic value for hotspot areas is 3 million IDR per m^2, with subsequent years valued at 40 thousand IDR per m^2. For non-hotspot areas, the year-one value is 1 million IDR per m^2, with the same annual value of 40 thousand IDR per m^2 in the following years. These functions and values are summarized in the following algorithm:

$$EV_{nct} = \left(HRC_t * L_t\right) \tag{8.4}$$

$$HRC_t = \left(3{,}000{,}000 * L_1\right) + \left(40{,}000 * L_t\right) \text{for the hotspot area} \tag{8.5}$$

$$HRC_t = \left(1{,}000{,}000 * L_1\right) + \left(40{,}000 * L_t \text{ for the non} - \text{hotspot area}\right) \tag{8.6}$$

where

EV_{nct} = Economic value of no-catch zone area
HRC_t = Habitat replacement cost for the selected area
L_1 = The extent area of the no-catch zone area in year 1
L_t = The extent area of the no-catch zone area in year 2 and so on

The values used to calculate HRC may vary depending on changes in labor, materials, and land prices. Nevertheless, due to limited data on price fluctuations in the study area, these variables were not included in our analysis. To account for inflation, we incorporated the consumer price index in the DSS's cost-benefit analysis. However, price fluctuations due to other circumstances have not been included. We recommend that future analysis includes these uncertainties.

8.2.6 Economic Valuation of Macroplastic Management

Macroplastic affects communities in several ways, including threats to human health as it degrades into microplastic, declining economic activities due to declining tourist visits, and increasing management costs to reduce its occurrence.

Ideally, economic valuation should be conducted on the outcome of microplastic management programs. However, we could not deliver the result of such valuation due to the mistake during the study design. Thus, we perform economic valuation of microplastic management instead.

We selected to study tourists' willingness to pay to support microplastic management programs because they are the stakeholder groups which own the financial capacity to support the programs, although local people are the most affected groups. The tourists, who are dominated by domestic origins, are divided into two groups: local tourists (originating from Solok and Tanah Datar Districts) and non-local tourists (traveling to the lake from other districts/provinces). Meanwhile, there are almost no foreign tourists recorded during the study time.

A pre-survey was conducted in February 2024 with 25 local and non-local tourist respondents to obtain the prioritized management actions. It resulted in four management priorities: (1) adding more trash bins around the lake area, (2) conducting regular macroplastic removal from the lake, (3) ensuring regular transport of collected waste to the dumpster, (4) promoting recycling initiatives.

Based on this result, a contingent valuation survey was conducted in May 2024 referring to the methods in [7]. The method allows for direct obtaining of people's willingness to pay (WTP) for improved macroplastic management. The survey was carried out in three tourism locations, representing differences in respondents and management types that comprise various response variabilities (Fig. 8.2)

After further screening, 110 respondents were interviewed, with 90 included in the final analysis (81.8% response rate). The applied criteria to elicit the respondents are their consistency in answering repeated questions and their completeness of the answers.

We used a tourism entrance for the tourism site as the bidding vehicle: occurrence of entrance fee in the currently free area (Darmaga Singkarak and Tanjung Mutiara) and additional ticket payment for the paid area (Tan Kayu). The principal questions of the surveys were:

1. Respondents' characteristics (education, gender, occupations, origins, education, and income).
2. Their awareness of Lake Singkarak's tourism and macroplastic problems.
3. Their opinions related to the macroplastic problems, whether and how it disturbs their tourism activities.

Fig. 8.2 Sampling sites for economic valuation of macroplastic management

Table 8.1 Average willingness to pay (WTP) for macroplastic management

Management type	Average WTP (2023 IDR/person)
Trash bins installment around the lake area	5806
Regular macroplastic removal from the lake	5250
Regular transport to the dumpster	5122
Recycle	5572

4. Their opinions about the microplastic management programs, based on the pre-survey results.
5. Their willingness to financially support the microplastic management programs.
6. Options for the amount of payment to support the programs.

The average WTP for each management program was calculated using their median value after being adjusted to remove the outlier data [8].

The results indicate that adding more trash bins generates the greatest WTP, followed by macroplastic recycling (Table 8.1). This suggests that respondents prioritize these two management actions over others. However, this finding contrasts with current management actions, which focus on incidental macroplastic removal from the lake (Jasman, 2002, personal interview).

8.3 Cost-Benefit Analysis

Cost-benefit analysis (CBA) is a tool to evaluate the economic feasibility of a project by assessing the stream of costs and benefits accrued during the project's timeframe [9]. It is selected due to its widespread acceptance as a standard appraisal method for public investment [10, 11]. CBA results are presented in two forms: deterministic Net Present Values (NPV) and benefit-cost ratio (B/C in the DSS outputs. In the default mode, the project expectancy duration is set to be 10 years. NPV and B/C are calculated using the following formula:

$$PV_B = \sum_{0}^{t} \frac{B_t}{(1+R)^t} \tag{8.5}$$

$$PV_C = \sum_{0}^{t} \frac{C_t}{(1+R)^t} \tag{8.6}$$

$$NPV = PV_B - PV_c \tag{8.7}$$

$$\frac{B}{C} = \frac{PV_B}{PV_C} \tag{8.8}$$

where

NPV=Net present value (IDR)
PV_B=Present value of the gross benefits (IDR)
PV_c = Present value of the costs (IDR)
R = Discount rate
B_t = Total gross benefits in year t
C_t = Total costs in year t
B/C=Benefit-cost ratio
t = Project duration (default: 10 years)

8.4 Sustainability Index (SI)

The calculation of a Sustainability Index (SI) is based on measuring the gap between the existing indicator values (Lake B in Fig. 8.3) and their ideal conditions (Lake A in Fig. 8.3), as expressed in Eq. 8.9. The SI was developed using stakeholder engagement methods to measure the lake's sustainability. The index comprises indicators derived from the components of the EBM framework. The weight of each component was determined using an analytical hierarchical process [12], based on the results related to each problem from the lake issues elucidated in Chap. 3.

The basic theory of the calculation is gap measurement between the existing indicator values (Lake B in Fig. 8.3) and the indicators under its ideal condition (Lake A in Fig. 8.3) (Eq. 8.9).

$$SI = \sum_{i}^{n} \frac{B(x_i) - A(x_i)}{A(x_i)} .k_i \tag{8.9}$$

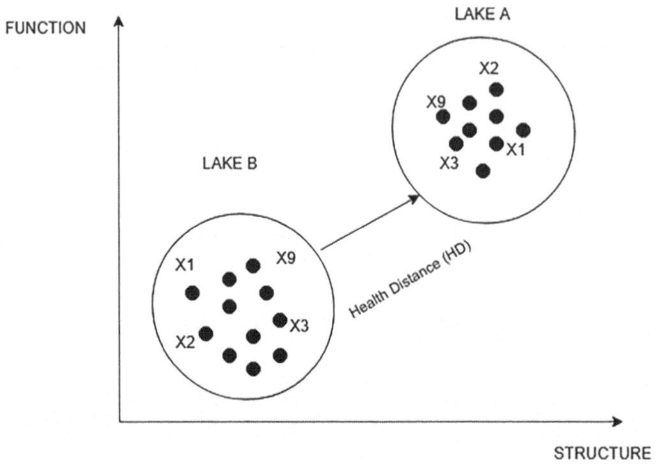

Fig. 8.3 Illustration of gaps between the current and ideal condition of the lake used as the basic theory for Sustainability Index (SI) calculations

where

$B(X_i)$ = the value of the existing indicators
$A(X_i)$ = the value of the ideal indicators
K_i = the weight of the indicators

The calculation results (Table 8.2) reveal that within the driver component, the type and number of fishing gear become the principal indicators (Table 8.2). Similarly, the number of illegal fishing gear with unsuitable mesh size is the most influential indicator in the pressure part, followed by the waste loading rate, particularly from macroplastic. Consequently, *ikan bilih* stock emerges as the most concerning component in the state part, while yield and fishers' income are the most significant indicators within the ecosystem services /impact components (Table 8.2).

Table 8.2 Weight and index weight of the components of EBM-DPSER framework (based on the AHP process reported in Chap. 3)

Framework components	Components	Indicators	Index weight (%)
Drivers	Agricultural activities	Agricultural land area	0.374
		Types of commodities	0.224
		Fertilizer use	0.150
	Domestic and tourism activities	Population	1.360
		Water use	0.816
		Disposal of household/ tourism waste	0.544
	Capture fisheries	Types of fish	1.016
		Number of fishermen	1.524
		Types and quantities of fishing gear	2.541
	Hydropower activities	Amount of electricity produced	0.766
		Water use	0.460
		Projected increase in electricity production demand	0.307
Pressure	Erosion and sedimentation	Erosion rate	0.935
		Sedimentation rate	0.401
	Nutrient loading	Input load TN	0.561
		Input load TP	0.240
	Unsustainable fishing	Number of fishing gear with illegal mesh size	5.835
		Number of fishermen who do not practice sustainable	3.501
		Number of fishermen and frequency of electrocution, use of poison, etc.	2.334
	Loading macro-microplastic and other contaminants	Input load of waste	4.573
		Input load of pesticides and antibiotics	1.960
	Alteration of hydrological patterns	Water level	0.766
		Current pattern	1.916
		Sluice gate opening frequency	1.149
	Alien species introduction	Number and frequency of input of tilapia and other foreign fish seeds	0.345

(continued)

Table 8.2 (continued)

Framework components	Components	Indicators	Index weight (%)
State	Water quality	Total suspended solid (TSS)	2.116
		Total nitrogen concentration (TN)	1.270
		Total phosphorus concentration (TP)	0.846
	Contaminants/ pollutants	Macroplastic density	5.571
		Pesticide and antibiotic concentrations	2.387
	Bilih fish stock	MSY/ fish stock	8.318
		Habitat area	4.991
		Nursery and spawning ground area	3.327
	Flooded rice field	Rice field flood incident	0.995
		The area of flooded rice fields	0.426
Ecosystems services	Clean water provisioning	Number of households using water	6.070
		Average clean water consumption per household	3.642
		Clean water prices	2.428
	Bilih fish provisioning	Number of catch per fishing gear and mesh size	10.516
		Fish price	6.310
		Fishermen's income	4.206
	Electricity provisioning	Electricity production	0.161
		Number of consumers	0.096
		Electricity prices	0.064
	Rice provisioning	Impacted area	0.829
		Duration of flood inundation	0.498
		Impact of floods on rice harvest and production	0.332

8.5 Conclusion

The equations and values reported in this chapter are the main integrating components that result in the outputs of the DSS. As the next step, we employed information technology (details in Chap. 9) to build the graphical user interface (GUI). We noticed that the economic value is sensitive to various factors that remain uncertain. This implies that it is crucial to carefully determine the model inputs and coefficients. We also noted that there is still homework to improve the economic valuation of impacts of macroplastic reduction on tourist visits and water quality. Therefore, we understand that the DSS development is an ongoing process, with the possibility of further adaptations as more information and feedback becomes available from the local government as the first and principal users of the DSS (further elaboration in Chap. 10). Overall, we highlight that being adaptive is one of the keys to the successful deployment of the DSS.

References

1. Kieslich M, Salles JM. Implementation context and science-policy interfaces: implications for the economic valuation of ecosystem services. Ecol Econ. 2021;179:106857. https://doi.org/10.1016/j.ecolecon.2020.106857.
2. Danelon AF, Augusto FG, Spolador HFS. Water resource quality effects on water treatment costs: an analysis for the Brazilian case. Ecol Econ. 2021;188:1–9. https://doi.org/10.1016/j.ecolecon.2021.107134.
3. Oh S, Rhodes J, Strong R. Impact of cost uncertainty on pricing decisions under risk aversion. Eur J Oper Res. 2016;253:144–53. https://doi.org/10.1016/j.ejor.2016.02.034.
4. Minister of Public Works and Housing. Penetapan Garis Sempadan Sungai dan Garis Sempadan Danau Dengan. Peraturan Menteri Pekerjaan Umum Dan Perumahan Rakyat Republik Indonesia NOMOR 28/PRT/M/2015, 28/PRT/M/2015 2015 pp. 10–7. Available: https://peraturan.bpk.go.id/Details/159992/permen-pupr-no-28prtm2015-tahun-2015
5. Strange EM, Allen PD, Beltman D, Lipton J, Mills D. The habitat-based replacement cost method. Fisheries (Bethesda). 2004;29:17–24. https://doi.org/10.1201/9780203486603.ch15.
6. Said DS, Lukman, Triyanto, Mayasari N, Nasution SH, Supranoto, et al. Domestikasi ikan pelangi Sulawesi Telmatherina ladigesi melalui aplikasi habitat buatan. 2007.
7. Boyle KJ. Contingent valuation in practice stated preference nonmarket valuation. The economics of non-market goods and resources a primer on nonmarket valuation; 2017. p. 83–130. https://doi.org/10.1007/978-94-007-7104-8.
8. Annisa TM, Harini R. Analisis Kesediaan Membayar (WTP) Untuk Mendukung Ekowisata Berkelanjutan Di Kawasan Wisata Gua Pindul, Kabupaten Gunungkidul. Jurnal Bumi Indonesia. 2017:1–10. Available: https://www.neliti.com/publications/228867/analisis-kesediaan-membayar-wtp-untuk-mendukung-ekowisata-berkelanjutan-di-kawas
9. Moosa IA. The costs and benefits of cost-benefit analysis as applied to environmental regulation. In: Handbook of environmental and sustainable finance; 2016. p. 407–28. https://doi.org/10.1016/B978-0-12-803615-0.00020-0.
10. Munasinghe M. Global climate change: economic and policy issues. In: Munasinghe M, editor. Global climate change: economic and policy issues. Washington, DC: The World Bank; 1995. Available: https://www.researchgate.net/publication/292798166_Applicability_of_techniques_of_cost-benefit_analysis_to_climate_change.
11. Addae-Dapaah K. Appraisal and cost-benefit analysis. In: International encyclopedia of housing and home. Elsevier; 2012. p. 70–5. https://doi.org/10.1016/B978-0-08-047163-1.00607-X.
12. Khan AU, Khan AU, Ali Y. Analytical Hierarchy Process (Ahp) and analytic network process methods and their applications: a twenty year review from 2000–2019. Int J Analyt Hierarchy Process. 2020;12:369–402. https://doi.org/10.13033/IJAHP.V12I3.822.

Chapter 9
The Connecting Information Technology

Satyo Gusti Anugrah, Ivana Yuniarti, Sigit Purnomo, and Djoko Budiyanto

Abstract The chapter presents the methods in the information technology (IT) part of the decision support system (DSS). As a final stage in the DSS development, it integrates all the aforementioned decision support tools to build a user-friendly graphical user interface (GUI). Starting with the elaboration of the data management and ending in the implementation and testing phase, the chapter guides the reader toward system building. Although a little IT background will ease the reader to understand this chapter, a general reader may also benefit, as we provide the definition and further elaboration on the more technical parts. We highlight that future DSS projects must provide enough time to build specific machine learning to enable more sophisticated features, such as simulations of uncertainty effects.

Keywords Decision support system · Information technology · Lake · Machine learning

9.1 Chapter Introduction

After determining the required inputs for the DSS and conducting the research to perform economic valuations (Chaps. 3, 4, 5, 6, 7 and 8), the obtained information and formulas are then processed as the inputs to present the graphical user interface (GUI). At this final stage, web-based software to connect all information generated from all previously described decision support tools (DST) was built using waterfall

S. G. Anugrah (✉) · S. Purnomo · D. Budiyanto
Department of Informatics, Faculty of Industrial Technology, Universitas Atma Jaya Yogyakarta, Yogyakarta, Indonesia

I. Yuniarti
Research Centre for Environmental Archaeology, Maritime Archaeology, and Cultural Sustainability, National Research and Innovation Agency (BRIN), Cibinong, Indonesia

K. Glenk et al. (eds.), *Decision Support Systems for Multisectoral Tropical Lake Management in Indonesia*, https://doi.org/10.1007/978-3-032-11268-2_9

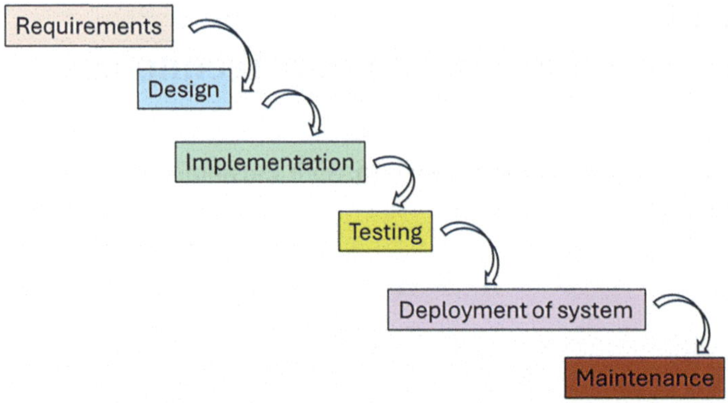

Fig. 9.1 System development phases in the waterfall method

software development methods. Waterfall is one of the sequential software development models in which each stage should be completed before proceeding to the next phase [1, 2]. This method is selected to provide definite structure and guidance in each development stage; thus, it can ensure the system's cohesion and quality [3].

The waterfall method has several phases: requirements, design, implementation, testing, system deployment, and maintenance (Fig. 9.1). This chapter elucidates the first four stages of the development system, which is systematically structured according to those phases. Further, we also provide some instances of coding followed by an analysis of the system's strengths and weaknesses.

9.2 Requirement Phase

The requirement phase is an essential stage where users' needs and understanding of the DSS are identified. The identified users in this phase are the research project team and the local stakeholders. A series of discussions with the project leader and selected stakeholders were arranged in 2023–2024.

From this phase, it is concluded that the built system should be able to incorporate both expert knowledge and data-based knowledge to quantify projected impacts generated by various DSTs. The system's core information integrates lake development's ecology and economic dimensions.

Data-driven information, which emerges with the surge of the availability of digital information, is considered another pertaining method to obtain scientific discovery [4]. Data-based knowledge dominates DSTs' development, as described in previous chapters. The information generated from each DST includes projected outputs, data, and algorithms and is then used as the system's inputs.

Despite a focus on data, expert knowledge [5] is a vital information source in the system's development. Expert knowledge is particularly useful in the quantification of sustainability index (see Chap. 8) since it aids in determining the weights of specific parameters, especially in the absence of relevant data or information. Although rooted in social science, the use of expert knowledge in ecological studies has been demonstrated by several studies, such as [6, 7]. At this stage, the project team interviewed stakeholders and several experts (See Chap. 3) to inform the development of the algorithm for the system's inputs.

9.3 Design Phase

Considering the requirements, the system was designed to transform the resulting algorithms into logic code. Moreover, to accommodate the multisectoral nature of Lake Singkarak's management, the system is designed using library technology and the Vue.js framework to create the front end. Meanwhile, to build the back end, we applied Node. Js. Express, and for the database, we used MySQL. Ultimately, the system is integrated into an integrated development environment using Visual Studio Code.

Vue.js is useful for building an interactive and responsive user interface, while Node Express is a light, fast, and high-performing server for running the algorithm. MySQL is a database management system that can aid in structured data saving and handle large databases and various queries. This technology can support an interactive user interface while granting easier data access and allowing data evaluation. Thus, it can adapt to future changes and further development.

9.3.1 Data Design

We used an entity relationship diagram (ERD) to visualize data relationships in the system (Fig. 9.2). This approach provides a structured presentation of information flows and data connections; thus, the connections between data can be easier to understand. The diagram also helps us to recheck the data when errors occur. In general, we categorized the data based on the traceback of each final output to categorize the input and intermediate input data. In total, we created 25 ERDs.

9.3.2 Architecture Design

9.3.2.1 System Overview

In the architecture design phase, we construct two principal parts of the system: the back end and front end. While the back end is the part where data is stored, processed, and managed, the front end is the part that directly connects to the users. The back end includes three main components: servers, databases, and application logic.

Fig. 9.2 An example of the entity relationship diagram

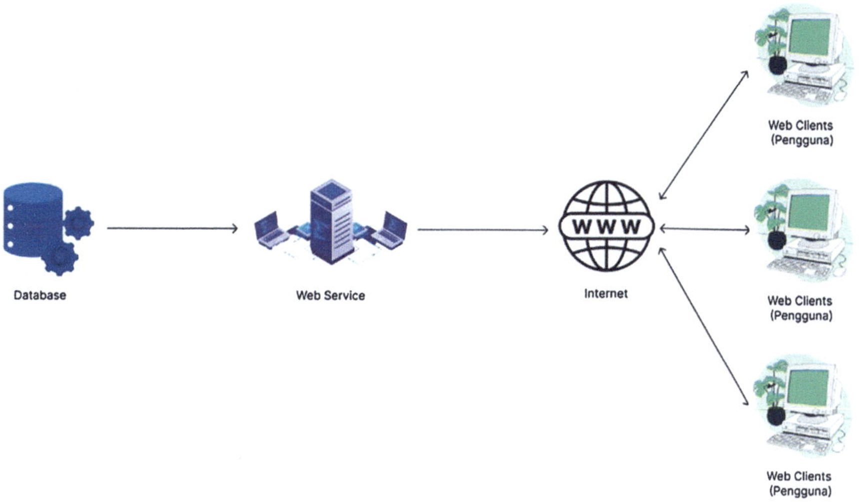

Fig. 9.3 System overview presenting how the database (in back end) is related to the front end to present the graphical user interface (GUI)

The front end comprises the technology to manage user interface, displays, navigation, and other interactive tools.

Both parts are then connected with a Representational State Transfer Application Programming Interface (RESTful API). The RESTful API is a communication method using HTTP protocol to enable users to interact with the server [8]. The user's request is processed as a request to the back end's provided API, which is then processed in the back end. This process will assist the back end in interacting with the MySQL database (Fig. 9.3). The next process is that the front end will proceed with the generated data by transforming it into Javascript Object Notation (JSON), which will be accepted by the back end and then being presented to the users (Fig. 9.3). This whole process ensures that the front and back ends can work separately to provide the services.

9.3.2.2 Software Architecture

The software architecture is the Model View Controller approach (MVC), which is a group of user interfaces presented based on user roles [9] (Fig. 9.4). Through the interfaces, users can access the provided features and functions to send requests to the back end. The MVC serves as a connecting bridge between user requests, generating suitable algorithms, and deciding how the data should be processed. The use of MVC guarantees that there is a definite separation between the interfaces, logic, and data manipulation. As a result, the system's modularity will be enabled; therefore, system maintenance can be easily performed.

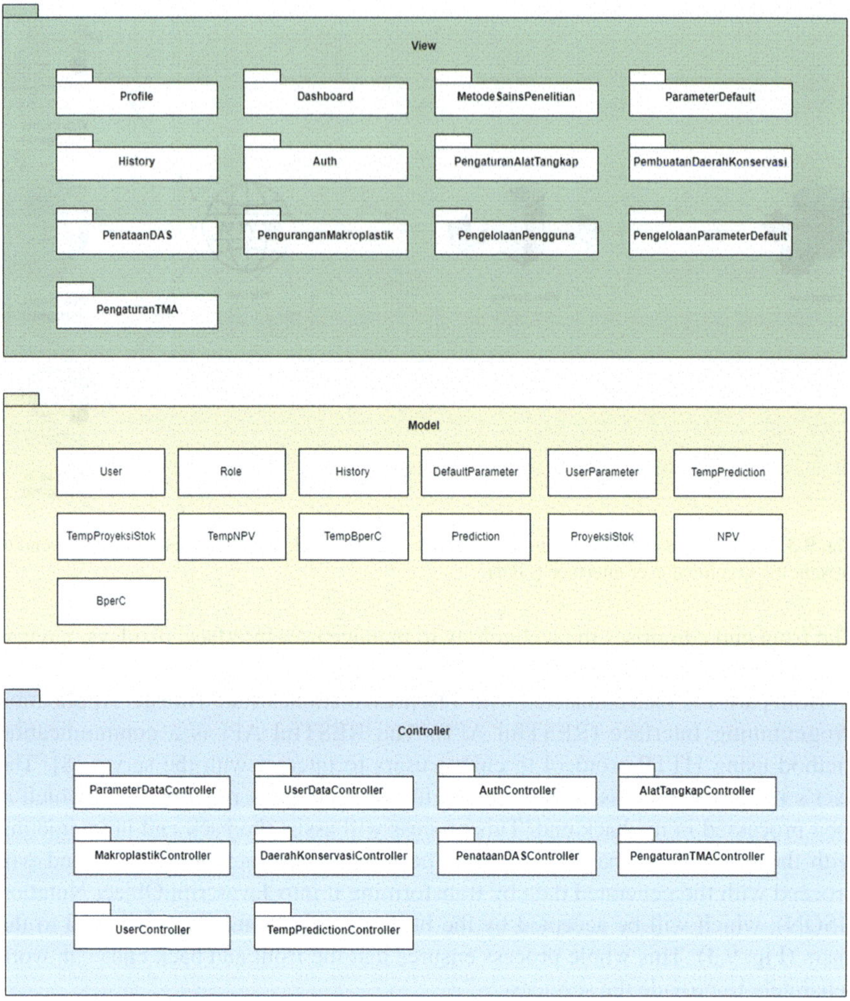

Fig. 9.4 Software architecture

9.4 Implementation Phase

9.4.1 System and Interface Implementation

The following code and interface examples (Figs. 9.5 and 9.6) illustrate the implementation of the design phase. The script code illustrated in Fig. 9.5a generates a command to conduct system checking on input fields via inputValidation () after the users click the login button. Then, the system will access the handlelogin () function in the API. It then executes the login if the verification in API is successful. If the

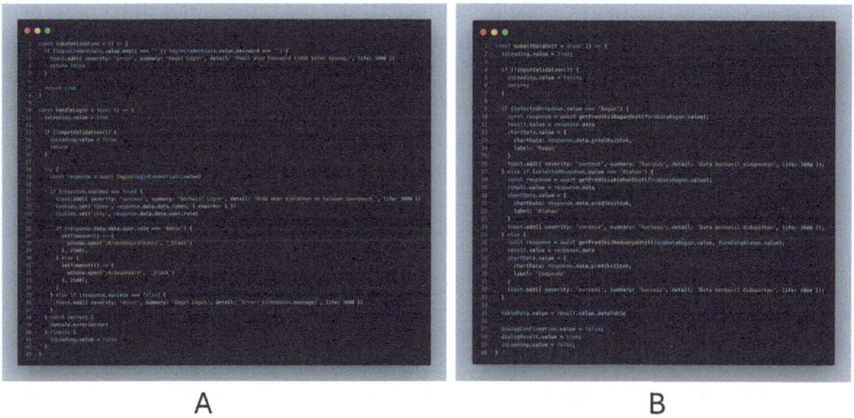

Fig. 9.5 An illustration of parts of code for (**a**) system checking on input fields via inputValidation for Login page, (**b**) verification for verifying all input fields in the fishing gear regulation page

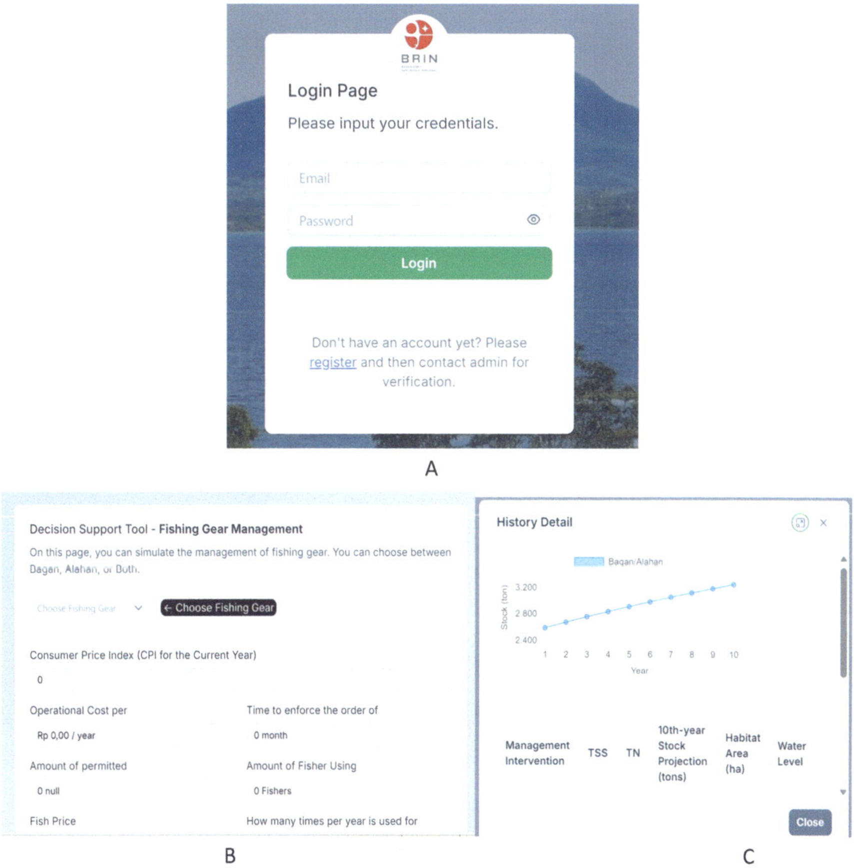

Fig. 9.6 Interfaces for (**a**) login page, (**b**) input field of fishing gear regulation page, (**c**) simulation results of fishing gear regulation

login is unsuccessful, an error test is conducted, and the error part will be presented to the users. Similarly, the code in Fig. 9.5b verifies all input fields in the fishing gear regulation page using the inputValidation () code. Next, the system will automatically check the user's elected fishing gear and access the PI via getPrediksiBaganUnit()or getPrediksiAlahanUnit(). Eventually, the system will receive the results from the back end, present the results *modal,* and then render the simulation graph and table.

9.5 Testing Phase

The functionality of the software is verified using the Black Box testing method [10]. This method focused on input and output functionality. The verification process results showed that they were well-functioned and valid (Table 9.1). The second verification test conducted with the stakeholders (the local government staff) also resulted in similar conclusions.

9.6 The System's Strengths and Weaknesses

As previously described, the software is free of charge, simple, and user-friendly. Furthermore, it can be evaluated and expanded to corroborate further enhancements to address future lake management issues. Moreover, using users' queries to accommodate data changes and limitations makes it flexible and adaptive.

However, there is still plenty of room for improvement in future DSS software development to anticipate further uncertainties. This system lacks specific machine learning programs to handle fluctuating data, such as price and real-time weather conditions. Further, an additional program to validate the user input data has not been built into this system; thus, we rely on other users' data and the user's manual comparison, which can be accessed via the history menu. These two limitations have become the software's principal weaknesses.

Table 9.1 Software evaluation results

Page	Description	Verification procedure	Input	Expected outputs	Resulted outputs	Conclusion
Login	User login page testing	Inputting email and password	Email:dummy@gmail.com - Password: dummy	Enable to access the DSS	Successful	Well-functioning and valid
		Pressing "login" button				
Register	Testing user registration	Inputting email and password	Email: newuser@gmail.com	Enable to register for the DSS	Successful	Well-functioning and valid
			Password: newuser			
		Pressing "register" button	User name: newuser			
Dashboard	Testing the dashboard interface	Login into dashboard		Dashboard and all features presented	Successful	Well-functioning and valid
	Testing the history of simulations of management interventions	Login into dashboard		Simulation history presented	Successful	Well-functioning and valid
	Testing the unsaved "compound" history	Login into dashboard		Unsaved compound history presented	Successful	Well-functioning and valid
	Testing the deletion of the unsaved "compound" history	Login into dashboard		Delete all unsaved compound notification	Successful	Well-functioning and valid
		Pressing "delete all" button				
	Testing of saving of all management interventions into "compound" mode	Login into dashboard		Saving into compound notification	Successful	Well-functioning and valid
		Pressing "saving all into compound" button				

(continued)

Table 9.1 (continued)

Page	Description	Verification procedure	Input	Expected outputs	Resulted outputs	Conclusion
Default setting management	Default overview testing	Pressing "default setting" button	Name: Waktu running	All current default is presented	Successful	Well-functioning and valid
	Testing adding parameters on default setting	Pressing "default setting" button	*Value:* 10	"Your default adding parameter has been successful" notification	Successful	Well-functioning and valid
		Pressing "add" button	Unit: Year			
		Inputting parameter name, value, and unit				
	Testing changing parameters on default setting	Pressing "default setting" button		"Your default change has been successful notification"	Successful	Well-functioning and valid
		Pressing the pencil icon in the parameter that wants to be changed				
		Inputting new data				
	Testing deleting parameters on default setting	Pressing "default setting" button		"Your parameter has been successfully deleted" notification	Successful	Well-functioning and valid
		Pressing the trash bin icon in the parameter that wants to be deleted				
History management	History management overview	Pressing the history management menu		All history is presented	Successful	Well-functioning and valid
	Testing deleting the selected history management	Pressing the history management menu		"Your history has been successfully deleted" notification	Successful	Well-functioning and valid
		Pressing the trash bin icon in the history that wants to be deleted				

Fishing gear regulation	Testing "single simulation"	Pressing fishing gear regulation menu	Input accordingly	Stock prediction graph	Successful	Well-functioning and valid
		Selecting the fishing gear option				
		Inputting information in the input field page				Well-functioning and valid
		Pressing "submit" button		NPV, B/C, and SI table	Successful	
		Pressing "single simulation" button				
	Testing "compound simulation"	Pressing fishing gear regulation menu	Input accordingly	"The compound simulation has been successfully saved" notification	Successful	Well-functioning and valid
		Selecting the fishing gear option				
		Inputting information in the input field page				
		Pressing "submit" button				Well-functioning and valid
		Pressing "compound" button				
Macroplastic management	Testing "single simulation"	Pressing macroplastic management menu	Input accordingly	NPV, B/C, and SI table	Successful	Well-functioning and valid
		Inputting data accordingly				
		Selecting the management option				
		Pressing "submit" button				
		Pressing "single simulation" button				
	Testing "compound simulation"	Pressing macroplastic management menu	Input accordingly	"The compound simulation has been successfully saved" notification	Successful	Well-functioning and valid
		Inputting data accordingly				
		Selecting the management option				
		Pressing "submit" button				
		Pressing "compound" button				

(continued)

Table 9.1 (continued)

Page	Description	Verification procedure	Input	Expected outputs	Resulted outputs	Conclusion
Water level management	Testing "single simulation"	Pressing water level management menu Inputting data accordingly Selecting the management option Pressing "submit" button Pressing "single simulation" button	Input accordingly	Habitat extent, water level, NPV, B/C, and SI table	Successful	Well-functioning and valid
	Testing "compound simulation"	Pressing water level management menu Inputting data accordingly Selecting the management option Pressing "submit" button Pressing "compound" button	Input accordingly	"The compound simulation has been successfully saved" notification	Successful	Well-functioning and valid
No-catch zone management	Testing "single simulation"	Pressing water level management menu Inputting data accordingly Pressing "submit" button Pressing "single simulation" button	Input accordingly	NPV, B/C, and SI table	Successful	Well-functioning and valid
	Testing "compound simulation"	Pressing water level management menu Inputting data accordingly Pressing "submit" button Pressing "compound" button	Input accordingly	"The compound simulation has been successfully saved" notification	Successful	Well-functioning and valid

Usage history	All simulation overview	Pressing history menu		All simulation data presented	Successful	Well-functioning and valid
	Testing simulation detail overview	Pressing history menu	No-catch zone	The simulation of no-catch zone presented	Successful	Well-functioning and valid
		Filling the search query				
Science and methodology	Science and method overview	Pressing science and methodology menu		Directed to the science and methods page	Successful	Well-functioning and valid
	Testing the overview on the details of science and methods	Pressing science and methodology menu		Methods pop-up is presented	Successful	Well-functioning and valid
		Pressing the selected methods				

References

1. Pargaonkar S. A comprehensive research analysis of Software Development Life Cycle (SDLC) Agile & Waterfall model advantages, disadvantages, and application suitability in software quality engineering. Int J Sci Res Publ. 2023;13:120–4. https://doi.org/10.29322/ijsrp.13.08.2023.p14015.
2. Mudassar S, Khan A. Waterfall model used in software development reference: software requirements engineering waterfall model. In: Software requirements engineering: Software requirements engineering practical approach; 2022. https://doi.org/10.13140/RG.2.2.29580.69764.
3. Murdiani D, Sobirin M. Perbandingan Metodologi Waterfall Dan RAD Dalam Pengembangan Sistem Informasi. JINTEKS (Jurnal Inform Teknol dan Sains). 2022;4:302–6. Available: http://www.jurnal.uts.ac.id/index.php/JINTEKS/article/view/2008
4. Savadjiev P, Reinhold C, Martin D, Forghani R. Knowledge based versus data based: a historical perspective on a continuum of methodologies for medical image analysis. Neuroimaging Clin N Am. 2020;30:401–15. https://doi.org/10.1016/j.nic.2020.06.002.
5. Hetmański M. Expert knowledge: its structure, functions and limits. Stud Hum. 2018;7:11–20. https://doi.org/10.2478/sh-2018-0014.
6. Costa A, Fenton N, Neil M. Integrating expert knowledge with data in Bayesian networks : preserving data-driven expectations when the expert variables remain unobserved. Expert Syst Appl. 2016;56:197–208. https://doi.org/10.1016/j.eswa.2016.02.050.
7. Rescher MD, Erera AHP, Ohnson CJJ, Use LJB, Rew CAD. Toward rigorous use of expert knowledge in ecological research. Ecosphere. 2013;4:1–26.
8. Maurya R. Application of restful APIs in IOT: a review. Int J Res Appl Sci Eng Technol. 2021;9:145–51. https://doi.org/10.22214/ijraset.2021.33013.
9. Basaruddin S, Fairuz M, Rauf A, Abdullah KA, Rahim M, Ridzuan T, et al. System for managing research and publication: model-view-controller approach. Turk J Comput Math Educ. 2021;12:5350–9.
10. Achmad YF, Yulfitri A. Pengujian Sistem Pendukung Keputusan Menggunakan Black Box Testing Studi Kasus E-Wisudawan Di Institus Sains Dan Teknologi Al-Kamal. J Ilmu Komput. 2020;5:42. https://doi.org/10.47007/KOMP.V5I01.4615.

Chapter 10
Lessons Learned and Future Strategies

Ivana Yuniarti, Klaus Glenk, Budi Kurniawan, Rosetyati Utami, S. Erni, Butar-Butar, and Hendro Wibowo

Some lessons can't be taught; they simply have to be learned.
(Jodi Picoult)

Abstract Reflecting on our journey in developing the decision support system (DSS) and our prior limited experience in its construction, we have identified several key lessons that can benefit future researchers aiming to develop DSS. The reflections are categorized into six areas: project timeframe and funding; thorough and adaptive planning; power of data and uncertainties; legitimacy and equity; participatory vs data-driven decision-making; and policy implications. At the end of the chapter, we illustrate further potential applications of the developed DSS in supporting the decision-making process in the study area. In the pre- and during the project timeframes, we emphasize the importance of professional and holistic research fund management, adaptive planning using an open framework approach, and the selection of professional and capable team members, which may be appropriate for all projects funded by the Global South Government. At the theoretical level, we reflected on the challenges to integrate participatory and evidence-based management and listed several important suggestions to improve their integration. We also contemplated the challenges and solutions to balance legitimacy and credibility in the DSS. We expect that our lessons learned will aid in improving future ecosystem decision-making-related projects.

I. Yuniarti (✉)
Research Centre for Environmental Archaeology, Maritime Archaeology, and Cultural Sustainability, National Research and Innovation Agency (BRIN), Cibinong, Indonesia
e-mail: ivana.yuniarti@brin.go.id

K. Glenk
Sustainable Ecosystems Team, School of Natural and Social Sciences, Scotland's Rural College (SRUC), Edinburgh, UK

B. Kurniawan · R. Utami · S. Erni · Butar-Butar · H. Wibowo
Research Centre for Limnology and Water Resources, National Research and Innovation Agency (BRIN), Cibinong, Indonesia

K. Glenk et al. (eds.), *Decision Support Systems for Multisectoral Tropical Lake Management in Indonesia*, https://doi.org/10.1007/978-3-032-11268-2_10

Keywords Decision support system (DSS) · Legitimacy · Lake · Decision-making · Credibility · Uncertainty · Policy · Participatory · Evidence-based · Lake management · Ecosystem

10.1 Project Timeframe and Funding

At the beginning of the project, three years seemed like a plausible timeframe. However, halfway through the project, we realized that building a holistic DSS requires significantly more time and effort than anticipated. We learned that longer (or more flexible) project timeframes are essential to address issues with the availability of verified and credible data. Not all data issues can be anticipated well at project start, and in our case, they hampered the construction of the decision support tools (DSTs), forcing the project team to allocate additional time to either collect primary data or to devise alternative plans.

Extra time is also important to cope with the complexity of the ecological system. Chap. 8 elucidates one example: we reflected on cases where we could not apply the preferred methodologies due to data unavailability. Significant factors, such as climate change, had to be excluded from our analysis because of a lack of secondary data and limited timeframes available to gather and integrate the necessary data.

Additionally, while knowledge co-production is vital, it demands extra time, resources, and funding, as reflected by [1]. Co-produced knowledge is regarded as a breakthrough approach to tackling wicked sustainability challenges, and it requires a process of iteration and collaboration [2, 3]. In our project, a team member was assigned to facilitate stakeholder engagement and produce the knowledge collaboratively. Nevertheless, the results were still lagging beyond what we initially expected. One key reason was the time constraint, compounded by the fact that most team members were still in the early stages of understanding how to make co-production work beyond simply presenting plans or results to stakeholders.

Unfortunately, most project funding in Indonesia, and likely in other Global South Countries, has not been designed to support multi-stakeholder discussions. Instead, they are often rigid and not tailored to specific research types. Most funding schemes prioritize laboratory-based research, focusing on laboratory equipment and materials, while providing little support for longer fieldwork and stakeholder engagement programs. Although separate funding schemes exist for stakeholder engagement programs, there are barely any connections between each funding scheme. This management of funding and resources poses a major challenge for the project team to attain expected and cohesive outcomes.

We recommend that Indonesia's project funding management adopt similar models engaged by international programs, such as the UK's Darwin Initiative, which offer a more balanced and effective approach. Funding scheme should balance the need to integrate ecological fieldwork, stakeholder engagement, and laboratory work, ensuring that all aspects of projects for sustainable management of the environment are adequately supported.

10.2 Thorough and Adaptive Planning

Having a contingency plan is crucial to guarantee success in any research project. In this section, we elaborate on the benefits of adaptive and comprehensive planning during project planning and implementation.

10.2.1 Pre-project Planning

The planning phase concerns selecting team members and project leaders, designing the funding proposal, and considering ideas and challenges, for example, through brainstorming. Although it is very useful to create a DSS project as an umbrella project for other smaller and basic research, if other projects are not co-designed together with the DSS project, it will be **an uphill battle** to embrace all other projects' outputs in the DSS due to differences in the way the data is collected and presented as reflected by [4]. Topic-wise, in our type of project, where multidisciplinary and stakeholder engagement are key, it has proven advantageous to approach the DSS as a stand-alone project rather than attempting to merge it with loosely connected research efforts.

Team member selection is pivotal for this type of project, especially given the multidisciplinary collaboration and the necessitated stakeholder engagement. Ideally, team members should be experienced, professional, diligent, communicative, willing to learn, and open to different perspectives (cf. [5, 6]). While involving new recruits may introduce fresh ideas, it may also come at an expense, such as delayed project delivery due to additional training needs [5]. In our experience, inexperience among team members contributed to difficulties in delivering output on time. Therefore, selecting team members with the appropriate qualifications can help avoid excessive time requirements that create issues for a tight budget [6–8]. However, we also wish to state that multidisciplinary projects with stakeholder engagement are excellent platforms for mutual learning; thus, added time and budget can be understood as an investment in future projects.

Further, contingency planning should be embedded in the project design, as the DSS project requires careful consideration of the high degree of interdependence among the project activities and with other external factors [9]. Risk identification, prioritization, and mitigation planning should be thoroughly undertaken during the pre-project stage and incorporated into the project proposal.

In our pre-project stage, we conducted a traditional risk mitigation planning based on expected risks and mitigation actions. We listed personal, teamwork, seasons, and stakeholders' risk sources and lined up conventional risk mitigation.

However, this approach was considered insufficient. Risks may arise from both technical and non-technical challenges. Conventional risk identification mostly focuses on technical challenges—such as staff health and weather conditions, while emphasizing non-technical challenges, including stakeholder disagreement and political conditions.

Thus, we altered our strategies by adopting three essential aspects of adaptive planning from [10]: *scenarios*, *options*, and *monitoring*. After extensive discussions, we decided to implement an open framework methodology proposed by [9]. This methodology allowed us to allocate buffer time and alternative methods, which turned out to be important to project success.

One of our biggest challenges was funding allocation. While most international funding sources have developed specific tools for risk management planning, research funding in Indonesia still operates in a more basic way, as hinted by [10, 11]. Effective risk management planning is recognized as a long-term investment, yet it frequently contrasts with the short-term nature of the project budget system [10]. Most projects are designed to be multi-year. Still, the uncertainty around public budgets—often influenced by political conditions— can disrupt funding, forcing projects to stop prematurely. We experienced this issue firsthand when our funding was cut in our first year, forcing us to condense a three-year project into a two-year scheme.

Learning from this experience, in special cases where the country's financial conditions are fluctuating, we recommend that the team divide a DSS project into multiple smaller sub-projects, perhaps each DST in each funding, and applying for several funding sources. It may make the DSS production take a longer time, but it will improve the quality of each DST.

10.2.2 During Project Implementation

No project is smooth sailing; thus, adaptability is one of the main requirements for successful project implementation, as also highlighted by [12]. We encountered several technical challenges, including insufficient and vague data, inapplicable methods, a lack of involvement from team members, and stakeholders' disengagement. We also experienced data discrepancies between different teams, which caused delays in the process. Eventually, it created an imbalance in the teamwork and hampered project delivery and meeting of deadlines. Ultimately, several important aspects had to be omitted to keep the project on schedule.

Applying an open framework during the pre-project phase was useful to mitigate the delay, as we had already allocated extra time, especially for data collection, DST development, and the graphic user interface (GUI). In one instance, we decided to allocate additional research on initial ikan bilih's absolute abundance, despite another team having claimed that they had conducted the research. By the end of the second year, we evaluated the other team's results, and we discovered that their data did not meet our expectations and that their methodology was flawed. Fortunately, we could have sufficient time to conduct the research, as we already have spared extra time and funds following the open framework suggestion.

However, not all external factors could be anticipated. During the project's first year, we experienced unexpected budget cuts due to the national COVID-19 reallocation scheme, which prevented us from conducting fieldwork for the entire year. To address this challenge, we adapted by conducting virtual meetings to gather stakeholders' perspectives and knowledge. Additionally, another research team provided financial support, enabling one of our team members to validate the results from these virtual discussions. Therefore, at the time of writing, we have yet to report on system deployment and maintenance outcomes.

From the experience, we recognized that although risk identification and mitigation have been performed, some factors remain beyond our control. As suggested by [13], our own limitations in dedicating sufficient time to risk identification may have also contributed to these challenges. For future DSS projects, we suggest maintaining adaptability and following an open framework methodology, as these strategies can significantly enhance project success.

10.3 Power of Data, Credibility, and Uncertainties

In this digitalized era, we postulate that data serves as the cornerstone of all environmental decision-making. As the heart of DSS development and output simulations, data empowers decision-makers to allocate funds effectively and prioritize programs. Therefore, ensuring data credibility is ultimately essential.

Here, we refer to [14] to define credibility as the scientific adequacy of the evidence and opinions used. It has become one of the backbones of evidence-based decision-making and DSS development. Without credible data, the reliability of evidence-based decision-making is doubtful, making it unsuitable for policy material.

Data credibility has become one of our major challenges during the project. While all primary data were rigorously assessed using scientific methods (see previous chapters), the input data from users and expert opinions may contain cognitive bias, often influenced by their personal interests and mono-disciplinarity training. We found that using expert and stakeholder weighing system in the sustainability index is the perfect example of the bias generated from mono-disciplinarity training. It was identified that most water quality experts assigned disproportionately high weights to water quality problems, even without reviewing the actual water quality data of the lake. Similarly, fisheries experts automatically referred to the decline of *ikan bilih* as the most urgent problem to be solved, regardless of broader ecological considerations. Learning from our first expert meeting, we experimented by presenting the data on both water quality and the *ikan bilih* fishery before the second expert meeting to reassess their weight assignments. The results showed that only a few experts adjusted their assessments after reviewing the data. Recognizing that bias persisted, we addressed this issue by inviting a more diverse group of experts with interdisciplinary backgrounds to subsequent meetings (third to fifth meetings) to reduce the bias.

This experience underscored the importance of maintaining a balance between single-disciplinary and interdisciplinary experts in decision-making panels. At the same time, we recognized that relying on data inputs from stakeholders, especially from those with vested interests in Lake Singkarak's management, may affect the credibility of future use of the DSS in the actual decision-making process. We had neither integrated a data verification system into the DSS nor assigned specific personnel to verify them. Instead, we relied on user discussions and critical thinking to manually verify their data and simulations. This case signifies the importance of establishing a verified and continuously updated public database for lakes, as hinted by [15].

Uncertainties emerge from various sources, including the nature of the projects, external factors, and the data itself. Specific uncertainties, such as water level fluctuations, have been incorporated into the model (Chap. 6). Data uncertainties are handled in each silo model (see previous chapters). The sources of uncertainty included in our analysis are:

1. The variability in initial fish stock data. The differences in the methodology used in secondary data sources led to substantial variations effects on the predicted stocks. To overcome this, we conducted a primary data survey using hydroacoustic (details in Chap. 6). After comparing the results, we provided two output scenarios based on primary and secondary data and relied on the stakeholders and expert panel decisions on the more plausible outputs.
2. Weather conditions:
   (a) Impact on water quality: Fluctuations in turbidity and other water quality parameters are incorporated into the habitat model by generating long-term water quality prediction scenarios in the mode development. After that, we compared the model outputs to obtain general patterns, ultimately, selecting the average values to build the final models as they can represent the value ranges (details in Chap. 6).
   (b) Impact on water level fluctuations: These uncertainties were integrated into both the habitat and flooded rice field models. For the habitat model, we simulated six water inundation levels, incrementing water levels by 0.5 meters between 361 and 363.5 m. We selected these ranges to reduce the spatial working loads. Meanwhile, to simulate the extent of the flooded rice field, we first calculated them for every 20 cm increment of water levels. However, after deriving the results, we did not obtain significant differences; we simulated them for every 50 cm increment of water level.
3. Economic conditions. The impact of economic shocks is reflected in price changes and the consumer price index in this DSS. The DSS allows users to input these two factors to provide the most updated information and ensure real-time adaptability.

4. Tourist preference. We included their preference on how microplastic management should be done and also the target of the management. However, we could only economically value the management, but not value the target outputs of the management due to several technical reasons (see Chap. 8).

Other significant uncertainties that can be important but are not included in the model construction are political instability, biogeochemical cycle alterations, sudden governance changes, social chaos, and climate change (Fig. 4.8). The latter highlights the need for integrating the DSS with the aforementioned database. Alternatively, future DSS updates could incorporate specific machine learning to account for evolving data uncertainties.

10.4 Balancing the DSS Credibility and Legitimacy

In this sub-section, we reflect on the challenges of balancing DSS credibility, legitimacy, and data equity. We have defined credibility as the scientific adequacy of the evidence and opinions used [14]. Meanwhile, legitimacy is related to stakeholders' acceptance of the data and the representation of their perspectives, values, and norms [14, 16]. Both aspects are essential in fostering the adoption of the DSS in practical decision-making.

To improve the DSS's legitimacy, we increased data transparency via stakeholders' discussions. We also included input data from stakeholders or users and engaged in stakeholders' discussions. However, this approach is risky, as the data collected from stakeholders may not be properly validated and verified, leading to credibility concerns, as explained in Sect. 10.3. Thus, we found that strengthening legitimacy and credibility can imply substantial trade-offs, as contemplated by [17, 18].

Conflicts between credibility and legitimacy became apparent during the framework construction and stakeholders' discussions. Most local people insist that the hydropower company's activities are the single and sole driver of declining water quality and fisheries degradation. However, long-term data series (***credibility***) emphasized that overfishing is the major driver of the fish stocks decline that opposed stakeholders' opinions (***legitimacy***). Although multiple discussions have been arranged, local people's perceptions remained largely unchanged, and more importantly, they questioned the research team's objective and degraded the team's legitimacy. This case underscores the fact that credibility and legitimacy frequently do not align, and in some cases, discussions alone cannot bridge the gap. We found that this experience corroborates the previous work of [19], which suggests that different institutional mechanisms are required to improve the legitimacy of the algorithm used in the decision-making process.

We suggest that the institutional mechanisms for future DSS projects include a pre-project stage assessment of details of stakeholder capacities, motives, interests, and how information is circulated between them. The assessment is imperative to reduce the trade-offs and conflicts between the credibility and legitimacy of the project's outputs. Further, we also advise that a long-term capacity-building program should be conducted improve how people understand data, data collection processes, and facilitate data interpretation. By integrating these strategies, future DSS implementations can better navigate the challenges of balancing credibility, legitimacy, and data equity.

10.5 Data Equity

Data equity, defined as the shared responsibility for fair data systems and processes [20, 21], becomes one of the main aspects of project implementation that the team and stakeholders should thoroughly contemplate. According to the data cycle framework explained by [20], data equity is characterized by two parameters:

1. Data sensitivity (privacy, regulations, cultural and commercial sensitivity).
2. Data accessibility (fairness, open access, digital inclusion).

At the time of writing, we are still in the process of formulating regulations on data privacy and the commercial use of DSS purposes. Meanwhile, we have made progress by opening access to the DSS and some data for the users. Access to the DSS is currently restricted to local and national government entities, referring to the requests of the district planning agencies. However, several case-sensitive data, such as respondents' credentials, remain confidential.

The DST development stage taught us a great lesson: equal data accessibility remains a major challenge in Indonesia. We acknowledge that even among our team members, we still collected and analyzed data in the DSS construction single-handedly, a ubiquitous situation in the research ecosystem in Indonesia. This time-consuming approach could be significantly shortened if equal access to verified data repositories had been granted. To improve data equity and accessibility, we propose the following strategies:

1. Establish an effective data-sharing system between research centers, universities, and other stakeholders. This highlights the urgency of creating a national lake database, which has been endorsed by the National Research and Innovation Agency (BRIN).
2. Increase adoption of ethical guidelines and research clearance in all research centers and universities.
3. Develop a standardized terminology to ensure consistency across disciplines and institutions.
4. Strengthen regulatory frameworks for data privacy and protection.
5. Conduct data governance and capacity-building training to enhance awareness and best practices.
6. Encourage joint research initiatives between research teams to foster collaboration and data sharing.

7. Establish a centralized data repository with a rigorous data verification process to ensure data quality.
8. Improve data governance and training programs to improve capacity building on data governance.

10.6 Participatory Vs. Data-Driven Decision-Making

This section reflects the extent to which stakeholders'information and data-based evidence are used. A widely cited argument in the social sciences asserts that "scientists need to learn to listen to the locals" [22]. There has been quite extensive literature highlighting the need for participatory decision-making in addressing wicked environmental issues (e.g., [23–26]). However, most natural scientists in Indonesia still hold to rigid evidence-based even though the evidence is partial.

In the context of DSS development and deployments, the remaining questions are:

1. To what extent can participatory data and information be included?
2. What kind of information should be considered?

We observed from our experience that in Indonesia and particularly in Lake Singkarak, local people, generally portrayed as the "masters of the ecosystem," may have a deep knowledge of their resources, but only to a certain extent. They can surely point out the production cycle, the season and weather changes, the fish movement, and the habitat spots. Their real-time observation can also be expected, compared to scientifically measured data. However, when asked about detailed information such as the fish migratory range and pattern of the fish reproductive season, it was almost impossible to receive certain and accurate answers. This indicates that while local knowledge is invaluable, it may not always be comprehensive or precise enough for DSS modeling.

Another clear example emerged during collecting of information about environmental change. We found that all local stakeholders, from the fishers to local governments, have uniform knowledge about what happened: everything was fine before the hydropower company came: abundant *ikan bilih* stock, crystal clear water, and no macroplastic deposition. This was particularly seen during the discussion session. However, individual interviews revealed differing perspectives; some of them raised the issue of liftnet operations as a key factor in the *ikan bilih* stock decline. If we delve deeper into the interview, we realize that they tend not to comprehend other factors, such as increased human settlement area, lifestyle changes, and the deployment of unsustainable fishing gear. This highlights a fundamental challenge: local knowledge can be too narrow and biased, because of personal and political interests, cultural upbringing, and economic motives. Perceptions may also be influenced by dominant voices within a community, leading to strategic misreporting or selective narratives. We know this finding may be unpopular amidst a surge in studies promoting local knowledge, and we generally support participatory

References

1. Beran D, Pesantes MA, Berghusen MC, Hennig BJ, Jacobi J, Lazo-Porras M, et al. Rethinking research processes to strengthen co-production in low and middle income countries. BMJ. 2021;372:1–4. https://doi.org/10.1136/bmj.m4785.
2. Schneider F, Tribaldos T, Adler C, (Oonsie) RB, De Bremond A, Buser T, et al. Co-production of knowledge and sustainability transformations: a strategic compass for global research networks. Environ Sustain. 2021;49:127–42. https://doi.org/10.1016/j.cosust.2021.04.007.
3. Norström AV, Cvitanovic C, Löf MF, West S, Wyborn C, Balvanera P, et al. Principles for knowledge co-production in sustainability research. Nat Sustain. 2020;3:189–90. https://doi.org/10.1038/s41893-019-0448-2.
4. Kleinsmann M, Valkenburg R. Barriers and enablers for creating shared understanding in co-design projects. Des Stud. 2008;29:369–86. https://doi.org/10.1016/j.destud.2008.03.003.
5. Chen SJG, Lin L. Modeling team member characteristics for the formation of a multifunctional team in concurrent engineering. IEEE Trans Eng Manag. 2004;51:111–24. https://doi.org/10.1109/TEM.2004.826011.
6. Cordero R, Farris GF, DiTomaso N. Technical professionals in cross-functional teams: their quality of work life. J Prod Innov Manag. 1998;15:550–63. https://doi.org/10.1111/1540-5885.1560550.
7. Feng B, Jiang ZZ, Fan ZP, Fu N. A method for member selection of cross-functional teams using the individual and collaborative performances. Eur J Oper Res. 2010;203:652–61. https://doi.org/10.1016/j.ejor.2009.08.017.
8. Chi YL, Chen CY. Project teaming: knowledge-intensive design for composing team members. Expert Syst Appl. 2009;36:9479–87. https://doi.org/10.1016/j.eswa.2008.12.015.
9. Zarghami SA. Project schedule contingency planning: building on von Bertalanffy's open systems theory and critical systems practice. Syst Res Behav Sci. 2024;41:247–61. https://doi.org/10.1002/sres.2965.
10. Malekpour S, Newig J. Putting adaptive planning into practice: a meta-analysis of current applications. Cities. 2020;106:102866. https://doi.org/10.1016/j.cities.2020.102866.
11. Jeuken A, Haasnoot M, Reeder T, Ward P. Lessons learnt from adaptation planning in four deltas and coastal cities. J Water Clim Chang. 2015;6:711–28. https://doi.org/10.2166/wcc.2014.141.
12. Deleris LA, Katircioglu K, Kapoor S, Lam R, Bagchi S, Adaptive project risk management. IEEE Int Conf Serv Oper Logist informatics. SOLI. 2007;2007:10598. https://doi.org/10.1109/SOLI.2007.4383874.
13. George C. The essence of risk identification in project risk management: an overview. Int J Sci Res. 2020;9:973–8. https://doi.org/10.21275/sr20215023033.
14. Cash DW, Clark WC, Alcock F, Dickson NM, Eckley N, Guston DH, et al. Knowledge systems for sustainable development. Proc Natl Acad Sci U S A. 2003;100:8086–91. https://doi.org/10.1073/pnas.1231332100.
15. Figueiredo MSN, Pereira AM. Managing knowledge—the importance of databases in the scientific production. Procedia Manuf. 2017;12:166–73. https://doi.org/10.1016/j.promfg.2017.08.021.
16. Vaara E, Aranda AM, Etchanchu H. Discursive legitimation: an integrative theoretical framework and agenda for future research. J Manage. 2024;50:2343–73. https://doi.org/10.1177/01492063241230511.
17. Heink U, Marquard E, Heubach K, Jax K, Kugel C, Neßhöver C, et al. Conceptualizing credibility, relevance and legitimacy for evaluating the effectiveness of science-policy interfaces: challenges and opportunities. Sci Public Policy. 2015;42:676–89. https://doi.org/10.1093/scipol/scu082.
18. Sarkki S, Niemelä J, Tinch R, van den Hove S, Watt A, Young J. Balancing credibility, relevance and legitimacy: a critical assessment of trade-offs in science-policy interfaces. Sci Public Policy. 2014;41:194–206. https://doi.org/10.1093/scipol/sct046.

19. Grimmelikhuijsen S, Meijer A. Legitimacy of algorithmic decision-making: six threats and the need for a calibrated institutional response. Perspect Public Manag Gov. 2022;5:232–42. https://doi.org/10.1093/ppmgov/gvac008.
20. World Economic Forum. Advancing data equity: an action-oriented framework. 2024. Available https://www3.weforum.org/docs/WEF_Advancing_Data_Equity_2024.pdf
21. World Economic Forum. Data equity concepts generative AI. 2023. Available: https://www3.weforum.org/docs/WEF_Data_Equity_Concepts_Generative_AI_2023.pdf
22. Reincke CM, Pieterman-Bos A, Bredenoord AL, van Mil MHW. Learning (how) to listen: a key aspect in training future scientists for meaningful dialogue with society. Int J Sci Educ, Part B. 2024:1–16. https://doi.org/10.1080/21548455.2024.2347531.
23. Robinson KF, Fuller AK, Stedman RC, Siemer WF, Decker DJ. Integration of social and ecological sciences for natural resource decision making: challenges and opportunities. Environ Manag. 2019;63:565–73. https://doi.org/10.1007/s00267-019-01141-2.
24. Reed MG, Abernethy P. Facilitating co-production of transdisciplinary knowledge for sustainability: working with Canadian biosphere reserve practitioners. Soc Nat Resour. 2018;31:39–56. https://doi.org/10.1080/08941920.2017.1383545.
25. Kaluma K, Umar BB. Outcomes of participatory fisheries management: an example from co-management in Zambia's Mweru-Luapula fishery. Heliyon. 2021;7:e06083. https://doi.org/10.1016/j.heliyon.2021.e06083.
26. Bunea A, Chrisp J. Reconciling participatory and evidence-based policymaking in the EU better regulation policy: mission (im)possible? J Eur Integr. 2023;45:729–50. https://doi.org/10.1080/07036337.2022.2144848.

Index

K. Glenk et al. (eds.), *Decision Support Systems for Multisectoral Tropical Lake Management in Indonesia*, https://doi.org/10.1007/978-3-032-11268-2

P

R

S

T

The manufacturer's authorised representative in the EU is Springer Nature Customer Service Centre GmbH, Europaplatz 3, 69115 Heidelberg, Germany. If you have any concerns regarding our products, please contact ProductSafety@springernature.com

Printed and bound by CPI Group (UK) Ltd, Croydon, CR0 4YY
05/10/2026
02228059-0005